DP

THE TRIBE

The inside story of Irish power and influence in US politics

Caitríona Perry

Gill Books

Gill Books
Hume Avenue
Park West
Dublin 12
www.gillbooks.ie
Gill Books is an imprint of M.H. Gill and Co.

978 07171 84828

Design and print origination by O'K Graphic Design, Dublin

Printed by CPI Group (UK) Ltd, Croydon CRO 4YY
This book is typeset in 12.5/17 pt Minion.

The paper used in this book comes from the wood pulp of managed forests. For every tree felled, at least one tree is planted, thereby renewing natural resources.

A CIP catalogue record for this book is available from the British Library.

5 4 3 2 1

For my husband and baby daughter

CONTENTS

EAST ROOM. THE WHITE HOUSE. 14 MARCH 2019.
6.12PM EASTERN TIME 11.12PM IRISH TIME

'This year, on March 17th, from Boston to Chicago, to the Emerald City of Seattle, and dozens of other cities and towns in between, millions of Americans will celebrate the legendary history and the rich heritage of the inspiring Irish people. I know many Irish people and they are inspiring. They're sharp, they're smart, they're great, and they are brutal enemies, right? So you have to keep them as your friend. Always keep them as your friend. You don't want to fight with the Irish. It's too tough. Too – it's too bloody.'

President Donald J. Trump

PREFACE

THE WELCOME'S ON THE MAT

There are seven times more Irish Americans in the US than there are Irish on the island of Ireland. One in ten people in the US claim Irish heritage, second only to those with Germanic ancestry. Unquestionably, Irish Americans form a big demographic. This has manifested itself for generations in the strong Irish influence exerted over American public life – from the local fire houses, police stations and city halls right up to Capitol Hill and on into the White House. Perhaps it's no surprise, then, that 22 of the 45 American presidents – on both sides of the political divide – claimed Irish ancestry. (The genealogy jury is still out on number 45, but with a Scots Gaelic mother, it is possible that President Donald J. Trump has some Irish blood.)

The earliest Irish immigrants created political strongholds; well-known names like the Kennedys in Massachusetts and the Daleys in Chicago. However, in the late sixties, Senator Robert Kennedy, himself in a race for the White House, began speaking of the death of the Irish American voting bloc. By 1968, Bobby Kennedy felt that Irish immigrants had so assimilated into American society that they were 'general' voters now, deciding on the issues in the round rather than Irish issues in particular. Of course, he did not live to see whether a veritable green army would carry him into the White House, as he was assassinated shortly after he made those remarks.

So why is it that a view still pervades in parts of the US, and in what feels like most of Ireland, that there is an Irish voting bloc?

Just how do the Irish in the US vote? And why is there a belief in Ireland that Irish Americans are mostly Democrats when, in fact, this is not the case? How is it that so many Irish in America voted for President Donald Trump when so many Irish in Ireland do not understand or like him? Has Irish America become more conservative than Ireland?

The key to understanding the Irish in America is to recognise that Irish America is not Ireland; it is, at best, a distant relative. It appears that the Irish in America – both Irish American and Irish born – are an assimilated, mature, immigrant population who vote as individuals, but who are bound together through a cellular-level connection to Ireland. In a nation as diverse as the United States, being Irish American offers an identity – a tribal binding.[*]

This emotional, spiritual instinct manifests itself in different ways. For some, it is embracing Irish dancing and music, for others it is an attraction to public service and social justice, and for yet others it is using political office, power and influence to connect to the Ireland of today in deference to the ancestors of yesterday. No matter the manifestation, Irish heritage gives a sense of pride, of belonging, of membership of a tribe that is older, bigger and stronger than the individual. But can that tribe mobilise into a powerful electoral force in the way that the other famous tribe – the Jewish in America – so ably have?

The Irish in America did not walk off a boat and into the White House, so this book will explore from the grassroots levels all the way up to the administration. What does it mean to be Irish for those in local politics, and for those who operate behind the

---

[*] This question of identity is reflected in the case of 'dropping the hyphen' when it comes to describing the possible dual loyalties of immigrant populations such as the Irish Americans. Modern usage favours leaving it out, as did several American presidents. 'There is no such thing as a hyphenated American who is a good American. The only man who is a good American is the man who is an American and nothing else,' said President Roosevelt. Similarly, according to President Woodrow Wilson, 'Any man who carries a hyphen about with him carries a dagger that he is ready to plunge into the vitals of this Republic whenever he gets ready.'

scenes, building networks and electoral influence? We'll hear from Republican and Democratic politicians in Massachusetts, including from the area that is home to the greatest percentage of people claiming Irish heritage – and which votes Republican. Many give their views on what it means to be Irish, from State Senator Patrick O'Connor to legendary Mayor Ray Flynn to the latest in the Kennedy dynasty – Congressman Joe Kennedy III – who doesn't rule out a tilt at the White House some day.

Does Irish America mean something else at the highest levels of politics in the US? What level of power can be wielded? Governors Terry McAuliffe and Martin O'Malley discuss. Perhaps of greatest strategic importance to Ireland is the influence it can exert in the United States Congress. Veteran congressmen Richie Neal (D) and Peter King (R) have different approaches, as does the only member with an Irish-born parent, Congressman Brendan Boyle. As for those working in presidential administrations, is there any consideration given to the Irish voting bloc, or is there just a party to be had on St Patrick's Day, with cheerful grins across a crystal bowl of shamrock? From the Trump administration, Chief of Staff Mick Mulvaney and former press secretary Sean Spicer weigh in, as does President Barack Obama's chief speechwriter, Cody Keenan. Former President Bill Clinton reminisces about the role of Irish America in his two presidential elections, explains just why he got involved with the peace process, and why he considers the Good Friday Agreement to be one of his crowning achievements. Past and present Irish and US ambassadors give a diplomatic view on the foreign policy relationship and the governmental perception of influence – and what it actually is.

All these politicians, presidential advisers, party operatives and community leaders from both sides of the aisle will not only outline what they think of Irish American politics and the influence Ireland enjoys in the US, but also where it stands to go in the future. Underpinning all of the contributions is a sense that

Irish America is shifting from a traditionally Democratic base, through the centre, and leaning towards the Republican Party and greater conservatism.

Changing demographics have a role to play in how powerful Irish Americans, and Ireland, can be. In 2016, for the first time in the United States, more Hispanic babies were born than white babies. Former US Congressman Joe Crowley, in office for 20 years, the son of an Irish-born mother, lost his seat in his New York district to the Democratic Party's rising star Alexandria Ocasio-Cortez in the 2018 midterms. He reflects on his career working for Irish causes, and why he was beaten so definitively by a young Hispanic woman.

The decades-long decline in the number of Irish people emigrating to the United States is also viewed as a threat to the future of Irish America and its influence. US politicians outline why they can't separate the case of the undocumented Irish in the US from those from other nations – the pain and isolation of living 'in the shadows' does not differentiate skin tones and neither can they. Republican and Democratic politicians explain why they're supporting the Irish government's push for more visas for Irish citizens. The Irish helped to build America, literally in terms of bridges and roads, and metaphorically. The Irish now account for only around 0.12 per cent of the Green Cards issued annually. But has 50 years of trickling Irish immigration already caused irreversible damage to the power of the community?

Brexit looms large on the island of Ireland, and has reared its head at the highest levels of the US political system too. That US congresspeople are threatening to block any post-Brexit US–UK trade deal because of allegiances to Ireland and the Northern Ireland peace process is a diplomatic 'coup', one that has left British authorities puzzled as to just how it happened.

––––

In the US media, the portrayal of Irish America, and the power it wields, is also enhanced by the Irish ties of many high-profile news anchors and talk show hosts.

Since then candidate and now President Donald Trump descended the gilded escalator of Trump Tower in June 2016, there has been much discussion of the Irish American heritage of the talk show hosts who support him – individuals who trade on being a 'tough Irish guy', like Bill O'Reilly and Sean Hannity. Many are household names and, although O'Reilly lost his main outlets due to allegations during the #metoo movement, still have a large conservative fanbase. Others are less well known, like Kimberley Guilfoyle, the former Fox News presenter, girlfriend of Donald Trump Jr and senior adviser on the Trump re-election campaign, whose father was born in Dublin.

This support has prompted many think pieces, like a 2017 article in *Newsweek*, which asks: 'Why are all Conservative loudmouths Irish American?'

There are popular Irish American conservative talk show hosts because there are conservative Irish Americans. But for every O'Reilly and Hannity, there is a Chris Matthews or Stephen Colbert, national network TV presenters who espouse liberal or Democratic viewpoints. Their arguments and actions are perhaps less sensational, and so garner less attention, but are none-the-less slanted to a specific political viewpoint.

Whether conservative or liberal, Republican or Democrat, the presence of many high-profile Irish American television presenters and radio hosts helps to get awareness for and coverage of Irish issues and affairs throughout the year, but particularly around the St Patrick's Day festival. Does this level of exposure to certain constituents help foster Irish influence? Does it make US politicians more likely to feel there may be an electoral advantage in becoming involved with certain causes?

This is not a history book. It is a look at the contemporary situation, with those currently involved, those who wield influence, and what they see and understand as the source and effectiveness of Irish power and influence in US politics. It is not intended to be an exhaustive encyclopaedia of powerful and influential Irish or Irish American politicians and political operatives. Nor can one volume stretch the length and breadth of the United States to include all geographic areas. Rather, this book is confined to some – not all – of the traditional strongholds to illustrate what it means to be Irish there, and how that results in political power and influence – or doesn't.

The powerful Irish in US politics are both Republicans and Democrats, and this book attempts to reflect that in those who are interviewed. However, both parties admit that Democrats have been more engaged with the community and so there are more high-profile Democratic players than Republicans. Operatives from both parties explain why.

The level of access that Irish people enjoy in the US can be surprising. Coming to work in the United States as an Irish person offers instant access by virtue of shared blood, shared ancestry, shared membership of a common tribe. It offers 'Brand Ireland' – a level of access that other countries can only dream of.

Power and influence are exerted by the Irish at all levels across US society. From the friendly barista who likes the Irish accent and offers a free extra shot of espresso, to the Irish cop who lets off the undocumented Irish person driving without a licence, to community activists honouring Ireland's history in their hometowns, to congresspeople, senators and presidents doing what they can to further and protect Ireland's welfare, to the unprecedented access to power in Washington DC that is enjoyed on that one special day a year.

But how long can it continue? Particularly in a United States that is becoming ever more polarised in its politics and divided in its society, where immigrant populations are growing and competing for access.

What does the future hold for all of this? What does waning Irish immigration mean? What does a prosperous island with relative peace need help with? What of Ireland in a post-Brexit world? Is the relationship more important to Ireland than to the US?

Irish America is not Ireland. The connections Irish America feels, the beliefs it holds, are not the same as those in Ireland. Many of those who feature in this book are intensely proud of their Irish heritage, even though they may not know exactly from where their ancestors hailed or when they came. They are proud of their Irish blood. They are tribal in their sense of belonging.

Ireland is the curator of the Irish experience. It's the touchstone for everything for an Irish American. To keep that going, connections need to be fostered with each new generation. The existential question facing Irish diplomats, in the interests of nourishing Irish power and influence in the United States, is how to rejuvenate and replace Irish America, particularly in the face of diminishing immigration. The challenge is how to marry the American understanding of Ireland with the Irish one.

AN IMMIGRANT'S TALE: FROM ARMAGH TO CAPITOL HILL

'Stop Inn' invites the blinking sign just visible under the Subway overpass on the corner of a crossroads in Woodside in the New York borough of Queens. From the outside, it just looks like any other big-city US diner – laminated signs stuck up in the windows, melamine pinned on the tables, wash-clean cushions on the seatbacks and curling paper notices taped to the doors about raffles, sports teams and charity drives.

The Stop Inn used to be known as Ray's when it opened in 1926. Old images on the wall show an Art Deco all-American diner and sign, which has long since been modernised, although there's greater romance in the old styling. The owner is Greek, the food is an American take on 'Irish' and the waiting staff are mostly Mexican.

While there's a sizeable Irish American community here, there is also a big Irish-born contingent, and the menu reflects that. You can upgrade to Irish bacon or Irish sausage from the standard American turkey fare for $1.75 extra. That's a fair deal. There is a coloured box on the laminate ten-page menu offering 'Irish Specialty', highlighting the cuisine that we've successfully exported down through the generations.

An Irish Special: Irish Sausages, Irish Bacon and Tea and Toast.

A Traditional Irish: Irish Bacon, Irish Sausage, Eggs, Grilled Tomatoes and Black and White Pudding.

Irish Mixed Grill: Lamp chop, Liver, Bacon, Sausage, Egg, Tomato and Pudding.

In a popular Dublin city diner, patrons may be seeking avocado toast, wheat pancakes and eggs Benedict rather than this fare, I suspect. However, the lunchtime menu here reflects many a lunchbox in Ireland, with an *Irish Salad Sandwich: Ham or Chicken with Lettuce, Tomato, Coleslaw and Cheese* also on offer.

But these appear alongside huevos rancheros and huevos Mexicana, reflecting in a tastebud-tingling way the changing demographics of the neighbourhood. As do the Saturday dinner specials: Irish Ham & Cabbage; Chicken Française; Spaghetti & Meatballs; Chicken Parmesan with Spaghetti; Beef Goulash with Noodles.

The Mexican waiter in his 50s – who doesn't want his name used – has worked here for 18 years. He says it used to be only Irish food that was served in Ray's/Stop Inn, and only Irish people who ate there. It was a virtual community hub at one time, and still is, judging by the posters taped to the doors – a fundraising drive for 'Roscommon Kids'; a Nua Eabhrac Pop-up Gaeltacht; an 'Irish Mile' (2km) road race and the upcoming St Pats for All parade. But the waiter says the Irish have largely moved on, and the menu has changed to have 'something for everyone'. There are still plenty of Irish customers, though, especially at a certain hour of the night once the Saints & Sinners (shamrock-festooned) bar across the road has closed. The Stop Inn is a 24-hour diner, and at a certain hour the Celtic Wrap is their best seller (Irish Bacon, Black Pudding and HP Sauce).

The snow is still falling in the heaviest storm of the winter so far. It's not fully daylight yet, so the fluorescent lights within are disproportionately inviting. The waiting staff at this early hour are all Mexican. The TVs are set to soccer – hard to find in many American diners, but not around here. Tottenham Hotspur are playing Arsenal and both sides are egged on by the diners through Spanish.

Joe Crowley looks around the diner wistfully. 'Here, in Woodside, I lost by 400 votes. My hometown, you know.' Although multiple people stop to say hello and commiserate with him, none stops long enough to explain why, after Crowley's 20 years as a United States congressman for the New York 14th district, the voters decided to plump for the firebrand newcomer Alexandria Ocasio-Cortez.

Much has been made about the decline of this Irish American politician. Crowley was in line to take over from Nancy Pelosi as the next Speaker of the House, but in a shock result in a 2018 Democratic primary, he lost to a fellow party member. The turnout was particularly low. Only 13.9% of those eligible turned out to vote. But a win is a win. Or, in this case, a loss is a loss. In a solidly Democratic district, the primary is more important than an election. Although it's minus 5 or 6 degrees Celsius on the morning we meet, the Hudson would have to freeze over before this immigrant community would vote for a Republican … but stranger things have happened.

There's no questioning Joe Crowley's Irish credentials. He is a first-generation Irish American on one side – his mother was born in Armagh and was brought to the US as a child. His father was born in the US, but both of his paternal grandparents were born in Ireland, his grandmother in County Cavan and his grandfather in County Louth.

'I used to say that when we came home from school it was like going back into County Cavan, once we opened the doors. We were always steeped in Irish history and music and culture.' He says he grew to appreciate that more through his teenage years and on into college and beyond. In fact, there is rarely an Irish party in Washington DC that is worthy of the descriptor 'party' that doesn't feature Joe Crowley joining the band for an Irish ballad or rock tune or two.

With such strong turn-of-the-twentieth-century Irish credentials, it is perhaps no surprise that big families and a tradition of public service run through the Crowley blood. His father was a police officer, as was his grandfather. His father was the eldest of seven children – 'depression babies' – and at that time, for immigrant families in New York, the police or fire department was a solid, pensionable job. But despite that NYPD Blue Blood, he was never pressured to join the force himself. But there was another strong Irish family business – politics.

'My father's brother – my uncle Walter – was in elected office. I remember when I was in the third grade, my uncle ran for the … I think it was the state assembly or the city council, and I had a bumper sticker on my school bag, with my name, and it was on a green bumper sticker, white lettering. To remind people that Crowley was Irish.'

Walter Crowley was a New York City councilman elected in this same area, Woodside, and when thinking back to those early campaigns with his uncle, Joe Crowley notes the area has changed.

'There was more of an Irish flavour, certainly, here at the time. More of a voting bloc. That would have been in the fifties and sixties and particularly in this neighbourhood in Woodside. It still has an Irish flavour to it, but as you can see walking around outside, it certainly has changed. I think we're the only Irish people here today! I think you'd have to really, outside the Irish bars, struggle to find a business owner here who's Irish.'

There's no question that alongside Boston and Chicago, and possibly Philadelphia, New York is identified as one of the biggest 'Irish' cities in the US. And although Joe Crowley is no longer a congressman, he has a legacy in the state of New York that will ensure the Irish American connection here is strong for generations. He is the reason that every student studying history in New York learns about the Great Famine. As a young man, elected to the state legislature, he was appalled to learn that New Yorkers didn't study – or understand – just why there were so many Irish in the Empire State. There was no mention of the reason 4.5 per cent New Yorkers claimed Irish heritage, and why they came in their masses in the late nineteenth century. As Crowley points out, what was 'done' (or not done) to the Irish that made the famine such a catastrophe – the treatment at the hands of the then ruling nation, the UK – was perhaps politically unpalatable.

'I sponsored a bill, a controversial one, but it required the teaching of the Great Hunger, or the famine in Ireland, in the New York State curriculum. And as much as it was about what happened to the Irish per se, it was also about the profound impact that it had on the development of the United States, and particularly New York. That drove a million people to America, many of whom changed the nature of New York. That was not written about, or barely mentioned in history books.'

His bill passed, and it's now required studying. 'If you take the Regents Exam[1] in New York State, more often than not there will be a question on the Irish Hunger. But it's also because of the use of hunger as a tool, a weapon, the subjugation that we've seen not only happen in Ireland, but in other parts of the world.'

Much was made of Joe Crowley's electoral decline in this New York immigrant district and that of Congressman Mike Capuano in Boston. Both middle-aged Irish Americans – in Capuano's case, the even more potent combo of Irish-Italian-American – who lost out to younger women of colour: Latina 'AOC' here in NYC, and African American Ayanna Pressley in Massachusetts.

Former mayor of Boston, Irish American Democrat Ray Flynn says Capuano was also good on Irish issues, but 'it's not the enclave

---

1 The Regents Examination is the High School Diploma awarded in the state of New York, the equivalent of the Leaving Certificate in Ireland.

of Irish voters like it used to be. The part of Boston where he represented is mostly black and Italian. These areas have changed.'

So, is it like the friendly waiter says here in Queens: the cycle of immigration has cycled on, and all the Irish Americans have just moved on? Not the case, says former Congressman Crowley. He also cautions against the narrative that a declining Irish presence means a declining Irish influence.

'The first misnomer is that this district was a huge Irish district. In the 20 years that I was in Congress, it was always a 70 per cent plus minority district, and so, roughly 30 per cent non-Hispanic white, which would be a combination of Irish, Italian, German, Polish, and other ethnic groups from Europe or thereabouts.'

Being the 'Irish guy' is not what used to get him elected here, he says.

'I've never had a predominantly Irish constituency, and so I think the longevity of the Irish in America was based not necessarily on our Irishness or being of Irish descent; it was a combination of things. In my earlier career, my state legislative days, there was more of a presence of the Irish in my district in a smaller, more concentrated way, but it was not the predominant vote. When I was originally elected, there was a larger concentration of Jewish voters in the other half of the assembly district. So the waning sense of Irishness in New York happened decades ago.'

Crowley says that although Irish America is diminishing in New York and elsewhere, the power of the brand is not necessarily declining. 'About 25 years ago, in Chicago, the Irish were no longer the ascendancy, but people with Irish last names were still getting elected. A lot of that has to do with the Irish being the least offensive to other communities. You know, if you were Italian, or Polish, you had the Catholic connection. For those of English or Scottish descent, well, if the candidates weren't English or Scottish, at least if they were Irish, they were close enough. I think the Irish have always been looked upon in America with some empathy and sympathy – it was a small country that had its own tragedies and its own difficulties, so people from countries with similar issues could relate.'

If he didn't lose his seat due to a declining Irish vote, what was the cause? He puts it down to some particular locations in his district – like this one, Woodside, and also the nearby Astoria and Sunnyside. Still anecdotally full of Irish, but the dive bars, small diners and walk-up townhouses have been replaced in the last two to three years by high-rise glass and steel boxes, juice bars, boutique concept gyms and vegan cafés. It's been gentrified.

It's a commuter district for Manhattan now; within a 15–20-minute subway ride to the island, in parts quite literally in the shadow of skyscrapers, but with rents that while not affordable by most standards, are considerably lower than Manhattan rates.

'I call it an "opportunity district". It's people coming from around the country, and around the world, looking for an opportunity to work. And they found it. They got it here in New York. They are talented people as well.' He says these newcomers were not as familiar with him, or with his work, as some of the more settled inhabitants were.

Changing populations are one thing, but Joe Crowley is well aware of the role that social media played. The bartender-turned-congresswoman Alexandria Ocasio-Cortez blitzed social media in an Obama-esque fashion, motivating voters and invigorating communities that may not have voted before. She continues to enjoy a high profile nationally, even if that means drawing the ire of not only some of the mainstream element of her own party, but also the wrath of the Republican Party.

He believes lots of voters simply didn't know who he was, but knew his opponent from her social media profile.

'I had never done television before in my history. I have never done paid-for ads in 32 elections I've contested, state assembly and in congress. But in this election, I did both cable and network television. And we spent well over one and a half, almost two million dollars on the primary.'

That's two million dollars for one candidate in one New York congressional district in a primary contest. None of it is refundable. That all goes into the coffers of the television stations.

'My hat's off to her, my opponent, Ocasio-Cortez. They ran a very effective campaign in terms of using digital technology and using IT to get that vote out. I'm not going to say that it was foreign to us, but we had not mastered it in the way that they had. It was by rote to them, whereas to me and to some of my team it was a learning curve.'

He says the same thing happened to Mike Capuano in Boston. His constituency was also an 'opportunity district'. Crowley says he lost too because there were a lot of millennials and young people in that district (comprising South Boston and the newly hipsterfied South Street Seaport area).

Crowley might say that his team had not 'mastered' social media in the way that the congresswoman with her own hashtag (#AOC) has, but he was labelled a dinosaur who was out of touch with that 'opportunity district' he describes.

He's aware of how he was branded. 'At 56, I was Grandpa Crowley. I thought – I'm 56 years old and I have a 13-year-old son. So I hope he doesn't have a child somewhere that would make me a grandfather.' There's a veneer of a chuckle. But you can tell it's been a tough road for the last few months. A change of life and direction is hard for anyone, at any age. But politics is a brutal game, and voters can never be taken for granted. Irish, Irish American or any kind.

Was he shocked by the result? Just then, a bulky African American man in some sort of security uniform passes the table. 'We miss you, Joe!', he shouts, and bumps his fist on the table as he passes. Crowley grins back and thanks the man, then continues. 'The shock of the loss? You can see it in some people around here,' gesturing towards the man and other patrons. 'I was definitely shocked. I don't think there's any question about it; it's not the greatest political upset in the history of mankind, but it was an upset. I don't want to diminish her victory for her because I think she did a good job. You know, it just …' he trails off, searching still to verbalise the sting of defeat.

Crowley says his family tradition of public service and his own commitment to having what he calls 'social consciousness' made the accusations in the wake of losing his seat last year all the more painful. 'I think this perception that somehow I was out of touch with my constituency or wasn't progressive enough really was not the case. It can be a little bit painful that it wasn't appreciated enough, you know, but that's what campaigns are about. And that's a lot on me. I didn't get that out enough ... which is painful, too. But that's life.'

And although no longer a member of the Friends of Ireland Caucus on Capitol Hill, the grouping of US senators and congresspeople with an interest in Ireland and Irish affairs, he has banded together with some other ex-congresspeople and senators to form an adhoc committee to protect the Good Friday Agreement and the peace process post-Brexit. There is perhaps little or no electoral benefit to him doing that now, so why all the effort?

'Just a genuine love of Ireland – to some degree maybe it is an infatuation. It's what we think Ireland was, what we think Ireland is, and what Ireland will be.'

Crowley hasn't ruled out running for elected office again ('I'll never say never'). For now, he's on the well-trodden (and lucrative) path from Capitol Hill across the National Mall in Washington DC to K Street – the home of 'consultants', or, as they are better known, lobbyists. But if he were to run again, he says, there would not be a coordinated effort to target any sort of existing or notional 'Irish vote'. He says that in the past, the Democratic National Committee, the party hierarchy, did target the Irish as a demographic voting bloc, but not any more. 'There was more in the past. They had an ethnic outreach, and they still do. A large part of that was Irish.'

But he says the issue of Brexit has potentially altered the interest that has waned in party HQ. 'It has kind of dawned on them; Brexit has awoken it again. I think there will be a movement towards getting people to take positions, especially with this president's lack of attention towards it.'

He's referring to President Trump's interest in Brexit beyond a trade deal. 'He's the first to assure Great Britain we'll do a deal, but it's not that easy.'

He thinks many Irish Americans voted for Donald Trump for financial reasons. 'There's an expression here – to vote with your pocketbook. And that's what Irish Americans did.'

But he says Ireland is not really a hot button issue. 'I think that there's a perception here that Ireland itself, and primarily the northern Irish situation, has been somewhat resolved. That's just not true. But what has changed that perception here a bit among the Irish American leadership is the Brexit issue and whether or not hard borders will come back. Economic pressures would have played a part, but it's the potential for the return of a border and British army outposts, and the tension that would bring, that has us rightly concerned.'

With diminished levels of immigration, though, he does think there is a risk that Irish influence might wane as more distance is created between the early immigrants and the country they left. But he has hope. 'Sometimes, people who trace their history back to the Great Hunger are ardent and strong and steeped in their own perception of Irish culture, and want to preserve and foster it through generations.'

Unlike many others, though, he is aware that there is a difference between Ireland as it is, and Irish America's perception of it. 'Modern Ireland is not the Ireland of our parents or grandparents.

But there's a nostalgia to it, a passion, a longing. It's not about the Emerald Isle or Claddagh rings or woollen sweaters and hats or music or art. It's about coming from a historical perspective, and Irish American elected officials have a political understanding that may be a little bit more sophisticated.'

His reasoning is more nuanced than many, where Claddagh rings, Aran sweaters, poor cousins in Ireland and conservative, religious family values fill their green-tinted dreams. And he is wise to that too.

'Some people here would think that everyone plays the Irish fiddle and sings Irish songs, or stepdances every day, but in Ireland, it's not the case. It's actually rarer than people here realise.'

But when it comes to wielding political influence, does the dream of Ireland matter more than the reality?

THE LOCALS

FROM GRASS ROOTS TO GREEN SHOOTS

Blink and you'll miss Marshfield Hills. Downtown has a grocery store-cum-post office (owned by Irish American actor Steve Carrell) and a school and lots of single-family homes – some with wild turkeys out front, others with boats in the driveways. And that's about it. It's the wealthier part of Marshfield, a beach town in southern Massachusetts with summer homes and caravan parks on the water's edge – some with their own moorings – many flying American flags out front, many flying Irish flags, some flying both on the same flagpole.

Marshfield Hills also happens to be home to the most Irish population in the United States – according to the 2017 US census, 63.8 per cent of the people here claim Irish blood, more per capita than in any other part of the United States. The title of 'most Irish county', by ancestry, jumps around several towns in this part of southern Massachusetts but right now it's claimed by Marshfield Hills. There's nothing surprising about Irish immigrants having settled in a rural community, not far from downtown Boston, on the sea, looking back across the Atlantic at Ireland, comprising low, rolling green hills and a topography reminiscent of the Home Country. What is perhaps surprising is that, when it comes to voting, this is a Republican stronghold.

Pat O'Connor is the local state senator. We meet in a seaside hotel. It's just coming into season, so the area is pretty quiet, although the smell of salt-and-vinegar chips and sugared doughnuts hangs in the air, giving the little town a sheen reminiscent of a cold winter's day in Bray or Salthill. The giant merry-go-round is boxed up, surrounded by temporary fencing in an effort to guard it from the harsh Massachusetts winters, keeping its bright colours well preserved for the summer that will inevitably come, no matter how heavy the snow or how frosty the air.

At the Nantasket Beach Resort tonight, there's a meeting to save the shoreline. The Save the Harbour, Save the Bay group is focused on keeping the water clean and preserving public investment in the beaches and coastline stretching from Boston Harbor to the South Shore. All the local politicians are in attendance, and it's a far cry from the Boston State House or even Capitol Hill. It is just like any community meeting in Ireland where the local county councillors turn up to show support. A look down the sign-in sheet shows the surnames are all Mc, Mac, O' and a host of other Irish names, while the faces inside the room also look vaguely familiar.

From here, O'Connor goes to a fundraising dinner and then has two other community events for the following two nights, putting in, he reckons, about thirteen hours a day. He's taking the coming Friday off, though, to see his family and 'grab a couple of drinks with my father'. But then he's back working Saturday and Sunday.

Patrick O'Connor was elected a Massachusetts state senator in 2016. Before that he was a councillor in Weymouth, Massachusetts, from 2006 until 2018. He was first elected to the council at the age of 21. He represents Plymouth and Norfolk, comprising much of the coastal districts south of Boston known as the South Shore, including the Marshfield region.

Although a Republican, he's been endorsed by unions representing teachers, police, firefighters, prison officers, drivers, bricklayers, carpenters, roofers, joiners and a whole host of other tradespeople perhaps considered to be traditional Democratic voters.

But in Marshfield Hills and the wider Plymouth County, the most Irish district in the US, the Republican Party is the dominant party. In the 2018 midterm elections, for example, where sitting Democratic senator and presidential hopeful Elizabeth Warren comfortably held her seat, gathering 60.4 per cent of the state-wide vote, the only county in the state of Massachusetts to vote for the Republican challenger Geoff Diehl ahead of her was this district.

Not only are they Republican voters, but President Donald Trump does well here too – Diehl worked on Trump's 2016 presidential campaign and is known locally as a 'Trump guy'. Trump lost Massachusetts quite dramatically in 2016, and here, in Plymouth County, Hillary Clinton still topped the poll, but he garnered 43.4 per cent of the vote, compared to the 32.8 per cent he got state-wide.

State Senator Patrick O'Connor calls himself Irish. His campaign colours on his logo and election campaign manifesto are navy and a shade of Kelly green. He's a little fuzzy on the actual ancestry, and says his Uncle Ed knows more than he does, but he knows both sides of his family trace their roots back to County Cork. His mother's family arrived from Ireland five generations ago, and on his father's side it's longer – he thinks probably about seven generations. But he is clear on their reason for leaving Ireland – they came to find the American Dream.

And he feels they found it. His mother is one of six children and his father is one of five – he jokes that much of the Irish family traditions are alive and well, namely having lots of children.

His mother, a nurse, grew up in Brooklyn, New York, and his father, a plumber, hailed from the well-heeled district of Cambridge, Massachusetts. They met, married and settled in Weymouth, MA, which O'Connor describes as 'predominantly a blue-collar Democrat community'. So although his parents weren't directly working in politics, they were in unionised occupations. His grandmother worked at one time as secretary for a politician whose name is spoken with almost the same reverence reserved for the Kennedys – Tip O'Neill, the Irish American Democratic congressman. She worked for him in the Massachusetts State House in the fifties, before he went to Washington DC, where he would eventually become Speaker of the US House of Representatives for ten years.

O'Neill's grandmother, Eunice Fullerton, was born just outside Buncrana, Co. Donegal. Throughout his political career, O'Neill worked to support the underprivileged in the US, using his position to push for peace in Northern Ireland and highlighting the situation at the highest US political levels. During the negotiations for the Anglo–Irish Agreement of 1985, Tip O'Neill promised British officials US financial support to underpin the accord. On the day it was signed, O'Neill and then president Ronald Reagan announced the establishment of the International Fund for Ireland. Since 1985, it has contributed millions of dollars to cross-border projects.[1] Tip O'Neill was granted honorary Irish citizenship in 1986, as was his wife. At that time, he was only the third person ever to have been so honoured.

Tip O'Neill's commitment to combat injustice was inspirational to his secretary, and State Senator Patrick O'Connor says his grandmother passed this spirit on to him. He says his work ethic came from his parents too. His father worked long hours as a plumber, and when he came home his mother would go to work nights in the local neonatal intensive care unit, taking care of premature babies. While she was gone at night, sometimes his father would be called to emergency jobs in Boston, and so would

1 Currently this fund gives $750,000 each year to cross-border projects. A resolution, with an amendment tabled by Congressman Brendan Boyle, the son of a Donegal emigrant, passed the House in mid-2019 and is awaiting Senate confirmation to double the fund to $1.5 million.

drop young Patrick off to his own mother's house in Cambridge, the matriarchal family home, to be looked after.

His grandmother would attend local county meetings – city council or school committees or whatever was happening. 'I saw it's important to be involved and engaged in government because it's probably the most direct way that you can provide assistance to people on a grand scale.'

He had caught the bug already and in high school ran for and was elected class officer (a sort of head-boy equivalent). He went to university in the UK, and when he was travelling over and back during holidays he began noticing the challenges that different communities faced and decided he wanted to be a public servant – 'to assist those that need help and bring government back to its core'.

He's been trying to do that from when he was first elected, but 'since I decided that this was something I wanted to get into – boy, has politics changed! It's become really divisive and not just in Massachusetts.'

While he may be a proud Irish American, he's also a proud Massachusetts man. He doesn't think things are good with national democracy, but he says they're trying to be a bipartisan state house in Massachusetts. This in spite of a heavily imbalanced political divide – the House of Representatives there has 127 Democrats and just 32 Republicans and one non-party member. The State Senate where O'Connor sits has 34 Democrats and he is one of just six Republicans. The US senators are both Democrats; all nine US congresspeople from the state are Democrats. But the governor, Charlie Baker, is Republican and in the 2016 presidential election 60 per cent of the voters in the state opted for the Democratic candidate, Hillary Clinton.

'I'm very proud that we're the domino that usually falls, that starts things. Whether it be equality or healthcare or public education – which was created in the Commonwealth of Massachusetts with

Horace Mann. A lot of Irish folks in Massachusetts take great pride in living in a state that proves that that sort of bipartisan public policy actually still works in America.'

Nevertheless, he doesn't get votes just because his name is Patrick O'Connor and his signage is bright green. 'Irish people in this district, they don't necessarily vote for you because you're Irish. There's a lot of ins and outs that come with it – they want to make sure that you're solid. And I think that comes from the upbringing that we have here, in a district of people who have Irish heritages – you know, you have to be a solid person. I think that they see through BS. They see people who aren't genuine. They want people to go up there, whether they're Republican or Democrat, and do the job that they were elected to do. And that's to represent people and to provide public services to those who need help.'

This comes across a lot from those involved in the Irish American community – outside a rigid party structure. Votes are gained based on reputation and earning the trust of the community, and, as a result, in recent years, the voters alternate between parties, voting for the candidate they like, not being faithful to a single party. Does that come from being an immigrant community? Does it come from those, in particular, who have been left with strong scars from the side they took in the Irish Civil War, when the political tribe you belonged to ripped families in half?

State Senator O'Connor says voter issues in his district are largely dictated by the Irish heritage of the majority of the people there. 'People here care about community-based services and helping people out. There's so much good going on in this district that I wish we could replicate all over this country. And a lot of that's from folks that have Irish last names and have Irish heritage. I think that's in our upbringing – from our grandparents and great-grandparents and all the generations that was passed on to – that you take care of people. There's a lot of what society is like over in Ireland still here.'

One of the towns in his district, Scituate, is a sister city of West Cork and O'Connor says he regularly gets to meet people from West Cork, and they have so much in common with each other and their communities.

Although Patrick grew up in a Democratic town, he is a Republican, but is quick to point out that he is 'a moderate, not a conservative'. That reflects the views of his district, he says. 'It's very middle-of-the-road folks. It's people who are open-minded. And I think that's one of the things that we are sorely missing now in society, open-mindedness. And people who admit mistakes and things like that. But I think you have it down here. And a lot of it's based off of heritage, but I think that's generally just the demeanour of people that live in the South Shore of Massachusetts, and why they call this home is because they're OK with voting on their ballot for Republicans and Democrats. Whereas in some areas of the state they just strictly vote for Republicans, or they just strictly vote for Democrats. But I've voted for many people and varying viewpoints. I think it's just illogical to hold one party in such high esteem that you will never vote for anyone from the other party. And I think that there's a lot of people who are like that in the district.'

Patrick O'Connor has been a registered Republican his whole life, but as he says, they are outnumbered in the state, and he hasn't always voted for fellow party members. But O'Connor doesn't think being outnumbered undermines the Republican Party in the state, nor that having a cordial relationship with their Democratic counterparts – rather than the polarised one seen at national level – weakens his party's position. 'It doesn't negate our influence on policy. Everyone, at the end of the day, knows that the thing that the people want the most is bipartisanship and they want this divide that's going on in the country to end, and they want people to come together and take a collective deep breath and say, all right, let's focus on the issues at hand.'

O'Connor's complaint is one made frequently by politicians at all levels, from both parties, blaming the opposition for a lack of progress.

'Health care's out of control. We have services that we need to increase for our disabled population. There are still more benefits we can give to our veterans. There are just so many things that, collectively, if we put our heads together and had conversations rather than arguments, we could actually solve, or get on the process of solving. But instead all we're doing on a national scale is having arguments and not really producing any results. I think that, hopefully, when the smoke clears, they can look towards Massachusetts and see there is a bipartisan government and it's working well. And it's working because it's focusing on shared priorities. And those shared priorities are public education, healthcare, veteran services, services for our disabled.'

O'Connor is correct in his expression of the efficacy of the bipartisanship of the Massachusetts legal system. For example, Republicans and Democrats in the State House of Representatives and the State Senate came together to pass a law that reverses funding cuts imposed by President Trump's administration to family planning clinics which offer abortion services. Not only did Republicans and Democrats vote for the bill that uses state funding to replace the withdrawn federal funding, but Republican Governor Charlie Baker signed it into law in April 2019.

Although President Trump lost the state of Massachusetts overall in the 2016 election, he was more popular in this most Irish district than he was elsewhere in the state. So how does Patrick O'Connor think the leader of his party will fare in the 2020 presidential election?

'I think that he'd probably do better here than in the state as a whole, but I don't think that he'd win. When we were out door-knocking and when we'd go into community meetings in the midterm election, voters were mad. People were voting for

the Democratic candidate, or voting against the Republican just
to send a message to the president. He definitely doesn't stand a
chance in Massachusetts, but I don't think that he'd win the district
either.'

Voters may flip-flop between parties, but here, at least at local
level, is one place where there is something of an Irish voting bloc.
It's not as cohesive a unit as it once was, but, O'Connor says, you
can win voters over by appealing to some common principles.

'If you give a compelling case to voters and you continue to do
good work when elected, then you can definitely get a good group
of folks, many of them Irish, who are so open-minded that they
don't care about the letter after your name, R or D. They just care
about what you're doing. But it doesn't hurt around here to have
an Irish name! I've been told often, "Oh, Patrick O'Connor is the
best name to have in this district"'.

But given that half the population here claims to be Irish, he
doesn't have a monopoly on Irish names. 'I serve with a James
Murphy from Weymouth and Patrick Kearney is in Scituate.' Both
colleagues are Democrats.

O'Connor says that, given that he is a Republican and he's not
from the traditional Irish-vote-getting party in Massachusetts, he
has to work extra hard. 'We have to go out there and prove our
worth, and prove ourselves to our constituents. And I think that a
lot of people with Irish heritage like the hustle, they like the hard
work. They like somebody going out there and grinding.'

And he is 'grinding' away as a local politician – the next election
is never far away and the need to fundraise huge amounts of money
never goes away either. He says a lot of the Irish in his district have
been very generous with their financial campaign contributions.

There are still active Irish organisations that can rally a vote for
certain candidates.

'The Ancient Order of Hibernians have a voting bloc in Scituate
that they definitely bring out. They work for and help candidates.

But I think that a lot of it is organised labour, and I've been blessed to have their support since the very beginning. I worked with them very well as a city councillor – they predominately endorse the Democrat, but they endorsed and worked real hard for me. A lot of the organised labour groups that we have in Massachusetts are very Irish. A lot of the Irish around here are in the trades.'

Despite the good fortune of his Irish surname, and his Irish donations, and his green-coloured campaign signs, O'Connor says he doesn't take the Irish vote for granted.

It's never, he says, a case of 'Oh, he's Irish so I'm going to vote for him. But it's the characteristics that Irish people like in individuals: hard work, determination, respecting elders, making sure that you're there for the right reasons. In Massachusetts politics, you've had a long line of Irish politicians who won even in times when their politics weren't popular. When Tip O'Neill won it wasn't popular to be a Democrat in Massachusetts. But he won because he knew that all politics was local, as his famous book is titled. And he knew what people wanted. Because he had grown up experiencing that, which is when a person's down and out, you lend them a hand and you try and help them out. When something needs fixing, you go out there and you put your best foot forward to try and fix it. And that's not a Republican and Democrat thing. And it's not only an Irish thing. But it's definitely a strong sense of the heritage you have that you look out for your neighbour and you make sure that, if they're down and out, you're out there trying to help them.'

He says there are some newer immigrant populations, more recent arrivals than the Irish, who do stick together and rally around candidates, but he says the Irish are not like that. Every vote has to be earned individually. The messaging might be the same, but every single Irish voter wants to feel like they've figured out what you're about and whether you're worthy of the vote.

For Patrick O'Connor, more than five generations from his Irish-born ancestors, being Irish is what drives him every day. 'A lot of who I am comes from my heritage and from Ireland. The Irish have contributed a lot to who we are as a country. I never forget that when I go up to the State House and cast votes and do what I do as an elected official. It's always there. I'm very proud of my Irish heritage.'

—

Across the state, inland, in Springfield, Massachusetts, there's a circle of Irish oak trees. They're little more than saplings now, but they will grow.

This is the only place in the United States that currently has a garden of remembrance dedicated to those killed in the 1916 Rising. It is also the only place outside Arlington Cemetery that has an eternal flame dedicated to President John F. Kennedy. These two memorials in Forest Park in Springfield are not a coincidence. It has one of the densest populations in the entire country of those claiming Irish heritage.

Pat Sullivan, the park ranger, says Springfield feels a 'special connection' to JFK, not just because he was a son of Massachusetts but also because he held a campaign rally in the city just before the 1960 presidential election. The eternal flame is closely tended to, as is the memorial, which is inscribed with one of the former president's most famous quotes – 'Ask not what your country can do for you, ask what you can do for your country.'

Forest Park is 735 acres. Pat Sullivan says the local people are 'pretty proud' of it, and he credits local politician, now a US congressman, Richie Neal with finding the money. When Neal was mayor of Springfield, he hired Sullivan and has been very helpful with getting federal grants for the park.

Ahead of the year of commemorations in 2016, the Irish government issued a call to action to diaspora communities to mark the centenary of the 1916 Rising.

Sullivan says that when that invitation was issued, they felt they had to build a garden. He says it's the fastest thing he ever worked on, and the community raised the funds needed – $100,000 – in just six weeks. Some people donated cash; some donated their time or the materials that were needed.

Forest Park was laid out by Frederick Law Olmstead, who designed a lot of US parks. In the middle, near the zoo, is a rose garden, which has been there since the park was built in 1880. The bottom of the rose garden was never developed, just grassed over, so it was deemed the perfect spot for a memorial garden.

Extremely careful planning was involved, and they stuck rigidly to the guidelines circulated by the Irish government. There are seven oak trees – upright columnar oaks that will grow to about 30–40 feet tall – one for each of the signatories of the Irish Proclamation, and a stone carving of the document itself. The stone slab was imported from Ireland and the text carved into it and then set into a larger slab that came from one of the original foundation walls of the city of Springfield. The centre of the garden is planted with boxwood in a Celtic knot, and around the edge are banners with the images and names of the seven signatories. The Irish tricolour flies day and night, lit by a solar-powered light after sunset to comply with the tricolour flag protocol.

So successful was the garden of remembrance, and so popular is it with the citizens, that ranger Pat Sullivan says they're now developing other memorials for the other ethnicities in the local population.

'We have the African American community, and they have a function called the Stone Souls. We're going to have a Stone Soul remembrance garden. We're working with the Hispanic community for a remembrance garden. It's kind of neat that it's bringing everyone's ethnicity forward.'

Springfield itself is similar to many New England towns of a reasonable size. It has a town square, ornate court building and city hall. The connections between Springfield and Ireland are woven throughout the town. Downtown, opposite one of many Dunkin' Donuts, is a sizeable tribute to one of the city's most famous sons, US Congressman Eddie Boland: a slightly larger-than-life bronze casting of Congressman Boland, accompanied by three stone slabs dedicated to his life story.

His memorial says that he was devoted to 'ideals which are the source of America's greatness', ideals that he was proud of as an Irish American, and he was an incredibly influential US congressman as a result. The influence of Irish America is visible not only in Springfield's past but also in its present.

Every year, in the run-up to St Patrick's Day in Springfield and the surrounding areas, there's a Colleen competition (lovely colleen not *cailín*). It's a Rose of Tralee-type competition and the winner gets to sit on a float in the parade and wave at the crowds.

Pat Sullivan's family has a good pedigree when it comes to this Colleen contest. He had three daughters crowned Colleen and his niece is the reigning Colleen for the nearby town of Westfield for 2019.

'We're all Irish here. It seems like all we hire here is Irish. When my grandparents moved over, they would tell you all about the signs in the windows: "Irish Need Not Apply".'

Pat says he remembers thinking everyone was Irish when he was growing up (probably because around here, they were). He didn't know any different.

'One day, someone was picking on me. I remember going in and crying about it and my mother says, "We're Irish. We don't care what anyone thinks about us. Go back out there and be proud you're Irish." Out I went. From that day on it was like, "I'm Irish.

Leave me alone!"' He laughs and beams with pride at the same time.

When he started school, racial integration in Massachusetts schools had not yet begun, so he went to the neighbourhood school with all the kids who lived beside him.

'Pretty much everyone in the school was Irish. Until I went to junior high, I thought everyone was just Irish. You'd have some kids that had Italian or Polish backgrounds, but only a few. My sisters have Irish husbands. My grandmother would ask anyone who walked into the house, "Where are you from?"'

His grandmother was from Cloghane at the foot of Mount Brandon in County Kerry. They had a large farm of about 400 acres. She was born in 1900 and came to the US when she was eighteen, following two sisters who had already made the journey and sent money back to help her out. She was sixteen during the Rising and, Pat says, told many stories about that period.

'Her story is the British soldiers would come through, and they would take pot-shots at her brothers when they were working in the fields and stuff. Then she told a story once how they came into the school and grabbed the rosary beads out of their hands and threw them against the wall, and the beads just shattered. She was concerned that we wouldn't forget about the Rising. She wanted us to know that it was so important for the start of Ireland's freedom.'

For Pat, to have been involved in the construction of a garden of remembrance for the Rising in his grandmother's adopted hometown has a level of meaning that, for him, is primordial. He has a tribal sense of pride at the achievement and also that he is the guardian of what he views as a sacred space. Forest Park itself is in a part of Springfield about twenty minutes' drive from where he and his grandmother lived. The spacious detached family homes here indicate it would not have been a district where impoverished immigrants first settled. 'My grandmother called this the Ritz section of the city. A few Irish lived here, but they were the lace curtain Irish' – middle-class rather than blue-collar Irish.

His grandmother married an Irish man, born in Coumeenoole in County Kerry. Pat says she went back to Ireland five or six times before she died, aged 98, but his grandfather, who died before Pat was born, never wanted to go back.

On his mother's side, the family were O'Sheas from the Blasket Islands and it was his great-grandfather Thomas O'Shea who left the Blaskets for America in the late 1880s. Settling first in Thompsonville, Connecticut, his grandfather then moved to nearby North Hampton, Massachusetts, and on to Springfield when he got a big job in a newly opened manufacturing plant. Both of his parents were born in Springfield and met and married there.

There is, without question, a strong Irish American community in Springfield to this day. The social scene centres on the John Boyle O'Reilly Club and the Irish Cultural Centre in nearby Westbrook Hill. John Boyle O'Reilly was a County Meath-born Boston journalist celebrated in this area for the contributions he made to Irish nationalism, the wellbeing of the Irish American community and the wider cause of American democracy. The club was named after him in 1880 when he was just 36 years old. A statue was also erected to him in Boston Fenway, the first Irish Catholic to have such a dedication in a city.

The change in how Irish Americans are voting is evident in Springfield. Pat says his parents, who've been Democrats their whole lives, are now 'not happy' with the Democratic Party because 'they've gone so far to the left'. His parents, both aged 89, have changed their voter registration as a result of their unhappiness and are now registered as independents. They haven't gone the full way to becoming registered Republicans, but they are pro-Trump. As are Pat's sisters.

Pat reckons more and more people will be leaving the Democratic Party and going independent.

Back downtown, we meet another man. He works for the city too and has been a long-time supporter of Democratic Congressman Richie Neal. He tells me his name but won't let me use it in print because he fears repercussions for what he tells me next. He voted for Donald Trump.

'I wouldn't lose my job or my friends if they knew because we respect each other. But I work for the mayor so there are certain things I have to shut my mouth on. Otherwise you just cause trouble for the administration. I keep my mouth shut and move on. Otherwise you would get branded, you'd be alienated.'

Just like Pat Sullivan, this man is a born and raised Democrat, his grandparents were born in Ireland and he is a proud Irish American. But for the first time, in 2016, he voted for a Republican, and that Republican was Donald Trump. And his reasons for doing so are similar to many of those in the so-called Trump 'base' – a group of people who will most likely see him re-elected in 2020.

'I want someone that's going to improve this country and make it strong. I don't see anything coming from the Democrats. Bernie Sanders? Well, he's a socialist. It's not good – it's not what we are in America.'

He says his parents and brother voted for Trump too. And although he says he's a Democrat, this man didn't vote for Obama either time – why? 'I just feel he had a lot of socialist views. They want to give it all away for free. And welfare? Oh, you don't have to work and we'll still pay you. That's not good. I think you have more respect for yourself and for your community if you're working actively in your community. I think you have a lot more problems when you're getting it all for free. Then you don't care. You're not productive citizens.'

Fairness is a big part of what drives this man and others like him. It is a concept Donald Trump needled during the 2016 campaign and will again in 2020. Why should one class of people do all the slaving and hard work, while others get to 'do nothing' and reap the rewards anyway? This feeds into the problem this city worker has with the presidential hopes of his local senator, Elizabeth Warren.

'Claiming she was Indian, I think, hurt her a lot because she's just a tiny fraction in her.'

He's referring to the controversy whereby Senator Warren said she was Native American, prompting Donald Trump to dispute that and use it to insult her, labelling her Pocahontas and engaging in other racial slurs. She took a DNA test in 2018 in an effort to prove her ancestry, but that angered members of the Native American tribes who view culture and kinship as the criteria for belonging to a tribe, not the potency of blood. She apologised to the tribe in early 2019. Senator Warren's DNA results showed that she had some genetic Native American traces from six to ten generations ago. While she has admitted informing officials at the University of Pennsylvania and Harvard that she had Native American heritage, there is no evidence that she ever used it for advancement. An investigation by PolitiFact found that her school records also showed that she had either identified as white or had not identified as a minority.[2]

This Springfield man thinks this episode has 'hurt her a lot' because of his point about fairness. Many white voters – and, in this region, that means many Irish voters – are for civil and equal rights and social justice, but they are not in favour of affirmative action. Their own immigrant background informs an 'each man/woman for themselves' attitude. The suggestion that Elizabeth Warren may have used affirmative action policies to get ahead is what is causing her some damage – even if the evidence is there that she did not benefit, the perception is that she did, and for some, that's enough. In this era of fake news that emanates from

2 Jon Greenberg, 'The facts behind Elizabeth Warren, her claimed Native American ties and Trump's 'Pocahontas' insult', politifact.com, 1 December 2017. <https://www.politifact.com/truth-o-meter/article/2017/dec/01/facts-behind-elizabeth-warren-and-her-native-ameri/>

the White House, where social media echo chambers are used as primary sources of information rather than traditional, factual journalism, myths and half-truths live on with the electorate.

He says, 'Again, it's about being truthful. You figure by her filling out those forms her whole life saying she was Indian, she got into colleges and it gave her a leg up. What would her life have been like if she had just said white? It's not good. She knew what she was doing, in my opinion. You just don't check that box off without having good data, you know? But again, she's just too far to the left for me.'

This fear about the Democrats sliding to the left is a concern for even the most staunch supporters within the party. They believe comments made by Bernie Sanders and Elizabeth Warren on the campaign trail and the high profile garnered by new congresswoman Alexandria Ocasio-Cortez could further damage the party. A caustic primary contest could ruin a candidate's chances before they even get to the general election.

This man is going to vote for Trump again, and there's only one person he would consider above President Trump: former Vice President Joe Biden.

'He just seems like a good guy, a good family guy. I have respect for him. I would vote for him over Trump. I'm happy with Trump but I won't say I care for the stories that come out about him as a person.'

The fact that the former vice president calls himself Irish and hails from another Irish American stronghold – Scranton, Pennsylvania – has nothing to do with it, the man says.

'I don't factor that in to how I vote – whether they're Irish or not. I look at what are they going to do for the common cause and are they going to make our community or our country great. I think Trump is turning – overseas, he's gotten, I think, more respect out of those other countries. It sounds like they're starting to pay their fair share. And it's going to be interesting to see how he does with China, with trade. That's questionable right now, but I think he's the right guy to make those deals – I think, before, everyone just kind of rolled over.'

President Trump's use of the media is something to be studied. This voter, a long-time Democrat, in listing his reasons for liking Donald Trump, sounds like he's repeating the talking points from a Trump campaign official prepped to go on a live network television debate, even using the nicknames that President Trump has used.

'The Clintons, they were just kind of crooked people, I think.' He voted for Bill Clinton in his first election in 1992, but not in 1996. 'He was too busy fooling around – he wasn't paying attention. He wasn't minding the store [i.e. the United States]. God bless Hillary, but what woman with self-respect would stay and allow her husband to behave like that? God bless her. It's their life and they do what they want to, but it's just not good.'

Despite the views of this man, Springfield is still a vehemently Democratic voting district, but it is changing.

———

Moving into the city, on a quiet cul-de-sac, perpendicular to the beach in South Boston, an older couple are keeping an eagle eye out for the bin truck. The city has just had its worst snowfall of the winter (12 to 14 inches in one night) and they want to make sure their barrel (wheelie bin) is collected. The streets around here are pretty narrow, but even under the heavy snowfall, the tops of deckchairs peek through in some of the parking spaces. South Boston – once the purview of the poor Irish – is now quite a diverse and hipster district, and parking spaces are at a premium. To make sure your spot is there when you get back from work, most people mark theirs out, not with a traffic cone, but with a deckchair.

The couple keeping an eye out are Ray and Kathy Flynn, for many years the first couple of Boston, as Ray Flynn was mayor of this town for nine years before being appointed US ambassador to the Holy See by then president Bill Clinton.

'The biggest issue we have here in South Boston is parking. It's like a civil war.'

Now 80, Flynn was first elected mayor in 1984, in what was then the biggest vote ever recorded in a Boston mayoral election. Italian American Thomas Menino replaced him in 1993 and he in turn was succeeded by another Irish American, Marty Walsh, in 2014. All Democrats. Since the 1884 election of Hugh O'Brien, every mayor of Boston has been Irish or Irish American, with the exception of Menino – and even he had an Irish son-in-law.

'Every one of these houses in the town had an American and an Irish flag hanging on the doorstep. That's not the case any more. Other ethnicities have moved in, certainly, and the Irish moved out. A lot of the Irish got more prominent. They got more money. They got better jobs. So now they're in the suburbs, which is not the centre of Irish activity. It's kind of scattered around.'

Ray Flynn says most of the early Irish immigrants lived and died in South Boston, but that's no longer the case. All his friends viewed moving to the suburbs as a sign of success, so that's what they did. But he stayed. Buying a house in the area now is unaffordable for many. 'This house here, a little house like this – it's about a million and a half dollars to buy. You can't afford to live here. It's just really unbelievable. So now it's professional people, well-educated people that work down in the legal and financial services sector in the seaport, they're the ones buying housing.'

He's a second-generation Irish American himself. His father's parents, the Flynns, came from An Spidéal in Galway, and his mother's parents, the Collinses, came from Woodfield near Clonakilty. It was the same village where Michael Collins was born but he doesn't think they're related. He still has relatives in Rosscarbery and Butlerstown.

Growing up, he says, there was no such thing as being aware of your Irish heritage. You were just 'Irish'.

'It was ingrained in you. The pride was just deep inside you. It was like St Patrick's Day was every day. We went through wars and we went through economic depression. We went through all the turmoil, but it never diminished the pride that the Irish had in themselves, their family and in each other. I'll never forget Mrs Crowley down the street, her husband got killed in the Second World War. I didn't understand it all that well because I was only five or six years old, but my mother made a big pot of beef stew and Mrs Ridge, next door, made three big loaves of Irish bread and Mrs Cane made some other kind of food, and we were all walking down the street, and my mother was carrying the big, big pot. And we did this two or three times a week for Mrs Crowley, in her grief. That's how they were.

'We had nothing. My father was in the hospital for six years – he got tuberculosis from working on the ships in the docks here. They said the highest rate of infant mortality of any community in America was right here in South Boston. And because of the planes, the ships, the contaminated harbour, we had the dirtiest jobs in America. And the factories, the big Edison factories – it was very unhealthy. Children died at a very early age. But despite all that, despite the fact that they had nothing, they had everything in terms of pride in one another. Commitment to helping one another. And even though they were Americans, they still had great pride in Ireland and talked about it all the time.'

For Ray Flynn, pride is the over-arching feeling and description of the Irish community that he grew up in and still lives in. He talks about seeing older immigrants 'walking around with the little shamrock on their lapel that they got 25 years ago from someplace. And that's what you miss.' He believes that as the immigrant community has become more successful, made more money and received a better education, they no longer need the Irish connection in the same way.

Flynn says when he first sought public office, there was no question among the electorate that he was anything other than 'the Irish Guy'. 'Voters pretty much knew that I was in the forefront for the cause of Irish justice and unity. I did this all year long, not just during the campaign trail. That's what we did, Kennedy [Congressman, then President], Teddy [Senator Kennedy] and the whole bunch of us.'

Flynn's mother's house was on the route of the St Patrick's Day parade and he remembers seeing then taoiseach of Ireland Éamon de Valera and then congressman John Fitzgerald Kennedy – or Jack, as Flynn calls him – walking in the parade.

'The people would go absolutely crazy and they'd follow them down the street. And there was no room to walk on the sidewalk. They were just mesmerised. I saw this as a kid and this had a big impact on me. And then, next thing you know, I'm throwing my little nickels in the big St Patrick's Day float up there to support some cause because it was the right thing to do.'

And in remembering the glory days, the regret returns, over the loss of an Irish power that no longer matches the fierce Irish pride. 'That level of pride and enthusiasm kind of dissipated, unfortunately. It was a fun period of time. I don't know how you get it back. I guess you can't. I don't know ...'

During his time in office, targeting Irish people in the city was extremely important. Then, three or four Irish newspapers were delivered to about 75 stores in South Boston, Charlestown, Dorchester, West Roxbury – the ethnic enclaves. He would send press releases on whatever he was doing to those newspapers in the hope of some valuable coverage.

'If you had a chance to go to Ireland, that certainly was a big event. When somebody from Ireland came over, or you could get the endorsement from some prominent Irish people or having your picture taken with the Irish people, you'd put them in those newspapers too'.

Flynn points out that Irish community newspapers have been hit by the same difficulties that have faced all local newspapers. Many no longer exist, others have been driven online. And local Irish community events are not something that the *Boston Globe* would have space for.

Ray Flynn says his success was due to taking issues he knew about because of his Irish background and translating them into common issues that would appeal to many people in Boston – those who were working class, blue-collar and had the same worries and concerns, regardless of where they came from.

Flynn says he his voting bloc was actually more of a coalition.

'The voting bloc was Irish immigrants, black people, and the unions. They saw my background, my mother cleaning office buildings at night, my father a dock worker and my wife Kathy's mother and father the same.'

He was invited to communities that identified with his background, but he says he didn't invent this sort of broad appeal. 'This was always a strong vote in Boston. It started with James Michael Curley [the son of Irish emigrants who fled the Great Famine, he went on to be a four-time mayor of Boston, the governor of Massachusetts and a four-term US congressman]. He fought back against the Yankee power structure, the financial power structure that ran the city at the time of the famous "no Irish need apply" signs. The people here always voted. The Irish felt that was an obligation as new Americans to vote. They'd a great pride in that, and a great pride in America.'

But the immigrant communities were also eager to be seen as American and to embrace their adopted homeland, so Ray Flynn says you couldn't talk only about Ireland and neglect American political issues – he had to stand on a broad platform. The 1983 mayoral contest was a defining one in the city.

'I ran against Mel King – he's a black fellow. My father was a dock worker, a longshoreman. Mel King, his father was a longshoreman.

The Irish, when they first came to America, they came to Boston and they worked as dockworkers. The ILA [International Longshoremen's Association, the union] was the base of power.'

As was the case then with many trade unions, at local level, and at national organising level, the Irish held the offices of power, and the sway of influence. It is often still the case now. Notwithstanding his union connection, the contest was fierce. 'We were seventeen points down. There was no way we were going to win. We had no money ... And we had no media support.'

One night, coming up to the election, both Ray Flynn and Mel King were taking part in a candidates' night, broadcast live on television – a big opportunity for them to impress voters. And for Flynn, crucially, it provided free access to a TV audience that he could not afford to pay for.

'Both of us got a standing ovation because we were talking about immigrants and poor people and social justice – by the way, we invented that term "social justice" here in Boston. And that resonated with Irish voters, with the immigrant voters that were Italian, Polish, Jewish. I was telling stories; I wasn't talking about, "we're going to bring this tax bill over here". I was talking about how I grew up in the Polish neighbourhood in Boston because there had been a major influx of Polish Jews at one stage in my neighbourhood.

'So, before you know it, I had every Polish person going out to vote for me. And the Jews. Mel King had all these immigrant black people, people coming from the south. And those were the issues, social and economic justice. Those were the issues of the day for everyone.'

Former Massachusetts state representative Mel King, now aged 90, was the first person of colour to make it to the final election for the mayor of Boston. His election-day face-off with Ray Flynn is the stuff of Boston political legend. Flynn, the grandson of Irish immigrants; King, the son of immigrants from Barbados and

Guyana. His mayoral candidacy rallied a coalition of black people and liberals, many of whom had felt estranged from the political process before then.

Ultimately, on election day Ray Flynn was victorious, winning 65 per cent of the votes cast – 82 per cent of the electorate had come out to vote in the biggest turnout in modern Boston history.

He got votes from across the city, and he believes his stance on education garnered him lots of them. He had 'Education is the answer' as a slogan on some of his campaign literature because of how well it played with the Irish (and other) immigrant populations.

'I got some money for a tutoring programme from one of the centres and we had about 60 or 70 kids going there after school so their mothers could work, and the mothers would come home from work and all their homework was done and the kids were getting great marks. And, these women would stop me in the supermarket and say, "The best thing that's ever happened to us since we came to this country is that tutoring programme." We didn't even think of it. But it took care of their kids, kept them off the street corners. It gave them a good shot at education. That's what the immigrants saw – particularly the Irish and the Polish – they saw that education is the key.'

Flynn is clear, though; he could not have won that hotly contested 1983 mayoral election – or his earlier city council elections – without the 'Irish vote'.

'No. Not at that time, and it was a decisive Irish vote. It was organised. I think I had a list of about seven or eight thousand Irish people alone who supported me.' Donating to political campaigns was not at the level it is now. Instead, his supporters would send out what were called 'dear friend' cards, urging everyone they knew to vote for him.

'It was a grassroots kind of organising,' says Flynn. 'It's different now. Everything is now television commercials. I don't know how you'd even organise the Irish in Boston now. Around what cause?'

And if Ray Flynn is saying this – a past master of the Boston Irish voting machine – then any Irish voting bloc is truly long gone.

The organisation was the key to everything. Ray says it was all about getting the vote out. 'I'd have four or five hundred people calling from the city of Boston. If they were at home changing their kids' diapers, they'd be calling the lady up the street, "Did you get out to vote yet or get somebody to pick you up?" That's the kind of organisation you had.

'Irish people recognised and understood the value of politics. They realised that education, being informed, being knowledgeable, getting involved, was the ladder in many respects to success, not just for them, but for their kids. We'd be encouraged by our family to get involved. They believed in the system. They believed in the candidates and the city itself – they trusted them.'

He doesn't believe people have that same level of trust in politicians any more, or the political system. The electorate just isn't motivated to vote.

'They'll go out to vote if it's convenient, but they don't have that fierce level of interest. Democrat, Republican, Irish or Catholic. It's not like that any more in America. It's special interest groups now.'

This was one of the major criticisms levelled at the Democratic Party following the 2016 campaign – that Hillary Clinton and her team in its Brooklyn headquarters were too interested in special interest groups and did not put enough effort into the issues Ray Flynn is talking about: jobs, education and local 'ordinary' working-class people trying to make a better life for themselves.

Ray Flynn is also a proud Catholic and was appointed by then president Bill Clinton to be the US ambassador to the Holy See from 1993 until 1997. While he doesn't think there is an Irish vote any more, he doesn't think there's a Catholic vote either. At least, not around Boston, where the Catholic community has been rocked by clerical sex abuse scandals.

'The people have become disillusioned. I don't know any Democratic elected office holder that faithfully supports the Catholic Church – abortion, marriage, nobody that supports those kinds of causes. The Irish people were very strong on moral issues. They're very strong on social justice issues for all people. Now it's kind of divided. We go to church two or three times a week, certainly every Saturday or Sunday. And you don't see the same kind of crowds in church like you once did.'

At one time there was an anti-Catholic backlash and Ray Flynn remembers being concerned about that when his good friend Jack Fitzgerald Kennedy was running for the White House. 'Kathy and I were with Jack the day before he got elected president of the United States. The question in our minds was: were all Protestant people going to go to the polls and vote against Jack Kennedy because he was Catholic? That was my concern.'

Ray Flynn and his wife saw JFK three times that day – in the Boston Garden, in the local school and in Providence College – and he says his only concern was over how his Catholicism would play out.

'We knew that he was a better candidate then Richard Nixon. But was there going to be a fierce anti-Catholic backlash, an anti-Irish Catholic backlash? Everybody was wondering whether he was going to carry the South, because they're mostly Presbyterians and Methodists, and the Protestant and the Anglo-American vote. That's why it was such a defining campaign for America. They broke down, but the Irish came out in fierce blocks.'

With several candidates with Irish backgrounds hoping to contest the 2020 presidential election for the Democratic Party, could the same level of influence be counted on?

'The Irish would not have the same kind of impact now in a presidential election. You might get 30 per cent Democrat and 30 per cent/35 per cent Republican. The rest of them are independent. They'll go to the polls and they'll say, "I don't know who the hell to vote for." They're not fiercely Democratic like they used to be.'

Flynn says the Irish voters in America are still 'fiercely conservative' or 'fiercely liberal', but don't fit the labels of Democrat and Republican in the traditional way. 'There could be 40 per cent or 35 per cent of the people going to the polls in November 2020 and they'll be walking down the street, and they won't know who they're voting for until they get to the polls. I could have told you in 1992 exactly who people walking down the street were voting for. They were voting for Bill Clinton. The Irish were a solidified group.'

—

Not far from where Mayor Flynn and his wife Kathy still live in South Boston is the Boston South Seaport. It's where he once worked, and both their fathers worked, and many of the blue-collar workers of the city worked. It's also where their grandfathers and many of the city's immigrants first landed in Boston.

It was once noisy and dirty and smelled of fish – Ray Flynn says he's never been able to eat fish, so much was his stomach turned by the experience of cleaning fish there. It was the chief port and wool, steel and anything that landed from Europe came through there.

But recent regeneration projects have rendered the area almost unrecognisable from those early days. As Flynn describes it, 'It's all prestigious law firms now. You'd need a hundred dollars just to get a couple of beers. Before, my father or Kathy's father could walk into an Irish pub with five dollars and get a few beers out of that, but five bucks would get you nowhere down there now.'

Five bucks would get you nowhere in pretty much any US city these days, but that's another story.

The area was mostly levelled and replaced with glass and steel office blocks and expensive apartments with rooftop pools, filled with young professionals, but there are still some shipping terminals.

In May 2017 the cruiseport terminal was named after Ray Flynn: the Flynn Cruiseport Boston at the Black Falcon Terminal. It was originally opened by then mayor Flynn in 1986 and welcomed thirteen ships in its first season. In 2017, when it was named after him, it was serviced by well over 50 ships and provided more than 950 direct jobs in the area and about half a billion dollars to the regional economy.

Beside the Flynn Cruiseport terminal is the Boston Marine Industrial Park. In February 2016 it was renamed the Raymond L. Flynn Marine Park to honour Flynn's legacy too. The official citation described Flynn as 'a native of South Boston with deep family ties to maritime commerce, [who] is widely credited with reviving the industrial port and preserving blue-collar jobs'.

Flynn says that at the official dedication ceremony of the Seaport terminal, attended by Republican Governor Charlie Baker, Democratic Mayor Marty Walsh and senators and other public representatives, he told many stories about Boston's immigrant history, about the people who had come through Boston port penniless in the 1940s and how their children had become doctors, nurses, teachers and journalists, that immigrants were 'the backbone of America'.

The issue of immigration in Boston has always been a live topic, from those earliest Irish arrivals through generations of Italians and Eastern Europeans. The current occupier of the office, Mayor Marty Walsh, has reaffirmed the city's status as a sanctuary city. Boston's Trust Act, passed in 2014, instructs Boston police to ignore detention and deportation requests from the Department of Homeland Security's Immigration and Customs Enforcement (ICE) agency unless the immigrants are wanted for a crime other than incorrect paperwork.

Mayor Walsh has been a vocal opponent of the effect the policies and commentary of President Donald Trump have had on illegal immigrants. Speaking in March 2017, during a visit by then

Taoiseach Enda Kenny to mark St Patrick's Day, Mayor Walsh said, 'I'm a proud son of Irish immigrants [both parents are Galway-born], but I would not be supportive of rules and regulations that just benefit people who are undocumented Irish. As mayor of Boston I have an obligation to represent all of the people in our city and that means all of the countries here. So, I couldn't support something that just benefited one country.'

Earlier that year, following President Trump's inauguration and threat to deport two billion illegal immigrants in his first week in office (which did not happen), Mayor Walsh offered sanctuary to all, holding a press conference at City Hall and speaking directly to those who were undocumented. 'You are safe in Boston. We will do everything lawful in our power to protect you. If necessary, we will use City Hall itself to shelter and protect anyone who is targeted unjustly.'

The circumstances were different, but Mayor Ray Flynn was doing the same in the 1980s. He remembers distinctly one incident around New Year 1989 involving two fishing trawlers – the *Laskara* and the *Kunataka* – that had arrived in the Boston Seaport from Poland.

'Poland was under communist rule at the time, and the ships had originally come out of Leningrad, but they had Polish sailors and sea merchants, and they got off the ship here at the Flynn terminal. Then they decided the crew was going to make a break for it, to get away from communism in Poland. They didn't know where to go – these were young Polish sailors, about 40 or so – and they were walking around and a cop went over to them and asked, "What are you looking for?" The cop was Polish. And the cop said to them, "Where are you from?" They finally, begrudgingly, told him that they'd jumped ship and they were looking for refuge. The cop says, "Well, you better be careful. The only place I could tell you to go is to see Mayor Flynn up at City Hall." So, they did. A whole bunch of them came up to City Hall.'

The sailors were all members of the Solidarity labour movement out of Gdansk, headed by Lech Walesa. Walesa would later become a good friend of Ray Flynn's and president of Poland.

Flynn continues the story. It was just 30 years ago but somehow feels like a lifetime, considering how the US has sometimes treated immigrants in recent years.

'They were in my office and after an hour, or two hours, word got around the city that there were these Polish sailors who had jumped ship – a communist ship. So, the INS, the Immigration and Naturalization Service, sent down five to six agents to get them out of City Hall and put them back on the ship. They were going to arrest them and ultimately send them back. The captain of the ship had no crew to get the ship back so he had taken out a complaint. The Immigration and Naturalization guy for the United States said to me, "You've got to give these sailors up – we're going to take them and we're going to put them in the wagons and take them down to the courthouse."'

This was not the right thing to say to Mayor Ray Flynn.

'So I said to him, "Now, how many immigration officials do you have with you?"

'"I dunno, five or six."

'"Well, let me tell you something. You know how many cops I have here in Boston? I have two thousand cops. Do you want your five or six guys to go up against my two thousand police officers? Because you're not taking these people out of my office."

'"Well, they're here illegally."

'"I don't care if they're here illegally. They're here for freedom, for justice."

'And he said, "Well, you're breaking the law, Mr Mayor."

'I said, "Well, I don't think I'm breaking the law. I'm certainly not breaking God's law. I know that for sure, and I'm sure you agree with me personally."'

'He starts laughing. He says, "I'd better get out of here. I'll go back and tell my authorities."

'And the agents surrounded City Hall with full force, but that was still only about 30 federal forces. And four days later we found a home for all these different people, and every person volunteered to take them. Ladies came down with food, offering rooms, and a Polish owner of the meat market was offering to help. Then there was the bartender of the bar room, he said, "We need people here. He can carry in all of the containers of beer or whatever it is." But that's what the big issues of the day were – helping immigrants, helping poor people. Because that was the Irish folks' roots. That's where they came from. Now they were in a little bit of position to help other immigrants and they did so.'

This was not always the case. There are plenty of examples of the landed Irish immigrants not helping the newer immigrants who came after them, keeping them out of certain accommodation and industries run by the Irish and the Catholic Church. There is evidence throughout US history of poor relations between Irish immigrants and African Americans, as well as between the Irish and other immigrants.[3] But despite this, Flynn feels, there is a social justice core to most of the Irish people who arrived into Boston. He laments what he sees as a modern-day diluting of community spirit.

'It was just a different time, different era. I don't know why it changed so quickly, but unfortunately it did. I suppose some

3 As outlined in great detail by the economist Thomas Sowell in his 2008 tome on ethnic America. 'Perhaps the worst relations between any two groups in American history have been between the Irish and the Negroes. Chronic animosity between them erupted into numerous fights and riots for more than a century, in cities across the country, both at work and at the slums they often shared. The famous draft riots of New York in 1863 saw rampaging Irishmen lynching Negroes on sight, often mutilating them, and even burning down an orphanage for black children … The Irish had similar relations with other groups whose skins were white. In addition to their many bloody clashes with the Scotch-Irish, they were also involved in numerous riots and street battles against the Germans; violence against the Italians in various cities, and attacks on Jewish property, persons and burial grounds in Boston and New York. On the West Coast, the Irish led both physical assaults and political attacks on the nineteenth century Chinese immigrants.' (Sowell, Thomas. 2008. *Ethnic America*. Hachette.)

people will say that I was wrong. It was the wrong outlook, the wrong policy. You sometimes bent the law. And I did, but it was for human rights and I'm sorry, but I'd do it again.'

—

Flynn sees the heyday of Irish political power and influence as being during the President Clinton era. When he was running for office and organising a national coalition for Bill Clinton, he remembers being a grand marshal of the St Patrick's Day Parade in New York City, and going to five or six events in one night. He thinks a unique set of circumstances and individuals at that time led to 'something special happening', but that the attention of the Irish Americans has fizzled out with the success of the peace process, and the Democratic Party has moved its focus. 'You see the Democratic Party are really going after the Latin American voters now.'

But why did the political benefit wane for American politicians? Flynn believes it's because the argument was won and done. 'There was no encouragement. There was very little media, very little praise given to any of these kinds of Irish American organisations. And then when we got the peace process, the Good Friday Agreement, people felt that, in many respects, the challenge was met, was over – we succeeded, we won. We had a cause to rally behind, to fight, to support, defend – that's gone.'

Many of the flagpoles flying Irish and US flags in South Boston may be gone, but as Flynn says, it's still one of the most Irish districts. But those left, as he says, have changed their registration to become independent voters or, dare he admit it, Republicans. And while Ray Flynn feels the Democratic Party has moved away from organising the Irish vote, Flynn notes Republicans moving into the vacuum left behind. The strong Irish Democratic machine kept Republicans out of certain districts, and without that operation in place, Republican candidates can get a hearing on the doorsteps.

'I think Democrats have concluded that that vote is not important any more, and so, as a result of that, every year it becomes less and less so.'

He laments the days when there were votes for all politicians who cared about Irish affairs, even those who weren't Irish themselves. He cites the case of the Republican US Congressman Ben Gilman, a Jewish man from New York who was in the House of Representatives for 30 years.

'Gillman says, "Ray, I never thought I'd be able to get the Irish vote the way I've been able to get the Irish vote. While I'm doing this, I'm supporting peace and justice in the United Ireland." He was there with Mario Biaggi [a ten-term Democratic Italian American US congressman from the Bronx who later went to prison on bribery and corruption charges] and all these different congressmen. It was the greatest thing because it was good for Ireland and good for the United States. It was particularly good for politicians, even non-Irish politicians, because the Irish voters would support any one of these people, Mario Biaggi, Ben Gilman, all these different Jewish, Irish, Catholic, Protestant, whatever, if they were committed to economic hope and opportunity and jobs in Northern Ireland.'

Nevertheless, at 80 years young, former mayor Ray Flynn and his beloved wife Kathy are still active members of the Irish American community.

'We go to all the Irish events here in Boston and sometimes I'll go to New York, but it's just not the same as it was.' He believes it is good politics to unite people and communities, and to keep young people involved by sharing stories with them.

He is sanguine about the changes that have come about in how the art or the game of politics is played. The old machine, the organisation, the face-to-face contact have all been replaced by mailing lists, Facebook posts and social media updates. The people are somewhat removed from the process.

But he says there is no longer a focus on current Irish news and affairs at many of the events he goes to. The people still talk about Ireland, but it's about the Ireland of a long time ago, during the Famine. That may be because restrictive immigration practices have quelled too much 'fresh blood' coming into the mix, and the stories that people know are those that have been handed down through the generations.

But mostly he blames the lack of the political machine, coupled with changing demographics – in particular, for the recent electoral defeat of a long-time local congressman.

Democratic Irish Italian American Mike Capuano lost his seat in the US House of Representatives in the 2018 midterms. His district covered much of the city of Boston, as well as outlying areas to the north and south of the city and the city of Somerville, where he had once been mayor. All areas that once had a high density of Irish American voters. There has been some redistricting over the years, but it is roughly the seat also held by JFK and Tip O'Neill. He was beaten in a Democratic primary challenge by an African American city councilwoman, Ayanna Pressley, who was more than twenty years his junior.

Ray Flynn gives his theory. 'His district has changed. It's not the enclave of Irish voters like it used to be. The part of Boston he represents is mostly black and Italian … The organisation is really something too. The organisation is not there.'

One area where Pressley scored much better than Capuano was Roxbury, which is now one of Boston's majority black neighbourhoods, but also home to the Irish Social Club of Boston. As Ray Flynn remembers it, it was also once a hotbed of Irish republicanism.

'When I was a kid, I used to go over to this Irish record shop in Roxbury with my grandfather. He fought in the West Cork

Brigade – we're a Collins family – and there'd be twenty, thirty members of the West Cork Brigade in the coffee shop having tea and scones and Irish bread. I was the only kid. And there'd be people lined up in the street – this is in the all-black area of Boston. They'd want to listen to their stories and other people would be coming by, taking pictures. And there was a fascination, a real romance about the history of violence and the Black and Tans and Irish culture and all the hardworking Irish men and women. There was a great deal of respect for it politically – it was extraordinary.'

Irish political influence is dead and gone, he feels – it has nowhere near the strength it had years ago. He describes the power as being a 'defining foundation issue', because the Irish were at the source, the foundation, of every part of American life. They were heavily involved in the unions, running civic affairs, organising civic events and even marshalling school matters.

'I went to South Boston High School, a public high school, and all the Irish ladies would be up there attending the parent–teacher meetings. And we had a big active parent–teacher organisation. The voting bloc of the Irish in America was really, really powerful and well organised.'

That secondary school – known as Southie High – was the scene of another famous Boston political face-off, one Ray Flynn remembers well.

'Ted Kennedy and Eddie McCormack had their first campaign debate up in South Boston High School. All the parent–teacher association women, all of a sudden, were up there and they got special seats over to the side.'

This was 1962, and a young Ted Kennedy was facing the two-term attorney general McCormack for the Senate seat that Teddy's big brother John had vacated when he became president of the United States. It was a seat the so-called Boston Irish political mafia were hoping to keep. And they did.

'The Irish were really active in the public high school, so they were active in the union, the schools, coaching – we had great sports programmes that they all got involved in – and the church. If you were a politician and wanted to succeed, you had to pay attention to that very, very well-organised constituency that was a voting constituency. I don't see that any more.'

And while the Irish may have lined the High School hall for the debate, an Irish newspaper threw up one of the most contentious points of the night. Although Ted Kennedy went on to win the seat, he fared quite badly during the debate when he was savaged by his opponent, who labelled him a parachutist. There had been much questioning of Ted Kennedy's Boston bona fides, as he had been raised in New York, London and Palm Beach and his father Joe had sent him on a tour of sixteen countries with the Senate Foreign Relations Committee to show that he was ready for the US Senate.

One of those countries was Ireland. When he arrived there, an article appeared in the *Sunday Independent* accusing him of being a carpetbagger. 'He's coming because later this year he is due to be involved in a political fight back in Massachusetts,' the piece read. 'He is playing the game political in what has now come to be known as the "Kennedy Method". You do not spend your time campaigning in your own little patch.'

McCormack read this article aloud during the Boston debate in what *Boston Magazine* describes as a 'gleeful' manner to 'a raucous crowd', delivering a blistering direct assault on his opponent, famously ending with the line, 'If your name was simply Edward Moore instead of Edward Moore Kennedy, your candidacy would be a joke.' McCormack finished by saying, 'You have never worked a day in your life.'

The morning after the debate, a smarting Ted Kennedy was out campaigning at about 6 am at a factory in East Boston. As the story goes, a big Irish American guy just coming off the night shift came up to Kennedy and said, 'I heard they gave you a tough time in the debate.' To which Kennedy agreed, and the man said: 'Let me tell you, Kennedy, I have worked every day of my life since I was fourteen, and you've missed nothing.' Ted Kennedy used to tell this story and joke that he knew at that moment that he would win the election. The Irish machine was clearly alive and well then.

Mayor Ray Flynn has dedicated much of his life to public service – for the people of the United States and for those in his ancestral homeland, and while he thinks Irish political influence is not what it once was, he's not giving up just yet.

'You got all this pride, you got all this knowledge and you've got all this love in your heart, but you don't know where to take it. It stays inside you. As long as I've got a little life in me, I'm going to continue to speak out for Ireland, but it's just not the same. You don't get the same response from the people that you got years and years ago.'

That is one of the biggest concerns, not just for US politicians who might be trying to rally a vote, but also for Irish diplomats and government officials who wish to keep the strong relationships and the connections as alive as possible in the corridors of power on Capitol Hill and in the White House. Is the Irish diaspora ageing out? Are the older folks still romantically connected to Ireland, through tales of poverty and fights for freedom heard from their grandparents, in a way that younger generations are not?

With his background knowledge of local elections in Boston and involvement state-wide and nationally in rallying an Irish vote – both for the Kennedy family and Bill Clinton – Ray Flynn is amazed that more local politicians aren't trying the same thing and haven't been interested in using his connections to keep it all alive.

'No politicians are calling me up and asking me for advice on how to get the Irish vote or what should we do? What should we put out? What should I say at this event? I think it's a mistake because I think there's great potential. The emotion and the pride are still there, but they just don't see it that way. And I'm wondering now, when I go, when my age group goes, is that the end of it? Because I don't see anybody coming in picking up the pieces.'

—

Martin O'Malley has been fighting elections for 25 years. First a Baltimore city councilman, then twice elected mayor of Baltimore, twice governor of Maryland. In 2016, he took a run at the White House when he unsuccessfully challenged former US Secretary of State and Senator Hillary Clinton and Independent Senator Bernie Sanders for the chance to be the Democratic Party's presidential nominee.

O'Malley is a proud Irish American who still regularly performs in an Irish rock band, and, although he wasn't born there, he's also fiercely proud of Baltimore. Having fought elections from the lowest level to the highest in a city that was once a blue-collar Irish immigrant stronghold, he has clear views on how much influence a common ancestry can wield.

Driving around the city, unable to find a particular CD to play, he bursts into song, 'Poor Paddy Works on the Railway,' an old folk song popular with Irish folk bands on both sides of the Atlantic.

In eighteen hundred and 41
My corduroy britches I put on
My corduroy britches I put on
To work upon the railway, the fucking railway
I'm weary of the railway
Poor Paddy works on the railway.

It's a song he often performs, and it tells the story of the immigrant Irish who worked on the American railroads, the first tracks for which were laid in Baltimore. The first stone for the Baltimore–Ohio Railroad was laid on 4 July 1828 by 91-year-old Irish American Charles Carroll of Carrollton, the last living signatory of the Declaration of Independence. The railroad went from Baltimore all the way out west to the Ohio river. It was the main means of transportation from the east to the west, and offered employment to countless newly arrived, and more settled, Irish immigrants – among many others. It was the biggest employer in Baltimore, and the city has poured money into a museum that honours that heritage.

Many of the surnames of those who built the railroad are still in the city today, their story told in a special tenement museum, just across the street from the Railroad Museum. The museum is in a building that O'Malley had protected from demolition when he was mayor of the city. On the street outside, there is a wall mural depicting the neighbourhood history. It is entitled 'Fáilte – Welcome' and it shows the African American children who now make up the local population playing ring a ring o' rosie with Irish children in turn-of-the-century clothing.

While the tenement museum tells the story of immigrant railroad worker James Feeley, the bricks on the street outside, bearing the names of the museum's donors, read like an Irish phone book – Barry, Carroll, Collins, Daly, Keaveney, Kelly, McNulty, McDonnell, McCabe, McManus, Murray, Murphy, O'Malley, O'Shea, Russell, Scally, Stokes, and so on. The fundraising for this museum continues, and in an ironic way, given the tone of the discourse around modern-day immigration under the current president of the United States. At the back of the museum, new donors can purchase a brick, inscribed with their name, which is then used to build a wall.

When he was first seeking elected office, Martin O'Malley was just 27, but he had a profile in the city from singing and playing music in Irish bars since he was in high school. He also knew a lot of people from his involvement in the annual St Patrick's Day parade committee.

O'Malley says he did his own form of data-mining to make a mailing list of Irish people.

'As much as the technology would allow at the time. I was running against an Italian American guy named John Anthony Pica, and so I certainly did a sort of the surnames in the district that were regularly voting Democratic. So I knew at the time what the percentage of Irish was in that district compared to the number of Italian surnames and Germans and the like, and I targeted them. At the time, it was considered high technology! I'm sure you could get a lot more precise about it now.'

He says it was really just a nucleus of friends and fans of his music, but he used them to campaign and knock on doors. The construction of a political machine, using the kind of retail politics that others lament has fallen off the agenda in recent times.

He knew where the Irish families lived, and knew that if he knocked on those doors, the people would be more likely to put up a campaign sign in their front yard or garden. 'Especially given that the signs were green and with an O apostrophe name!' he chuckles.

And in the greatest tradition – it was a family affair. 'My little brother ran the campaign, which is of course the Irish American way, right? He didn't choose to, he was just pressed into it by his pecking order in the family. I would come home from knocking on doors and one of the first things he'd say was, "Five people tonight were yeses and they all heard you play in the band."'

And although he courted what Irish he could and had great electoral success, at least until the Democratic presidential primary, O'Malley feels the impact of common ancestry varies. 'It probably depends on the geography. It's certainly a secondary thing, but more likely a tertiary or even lower priority.'

He believes the higher the office, the less likely people are to vote for a candidate just because they are Irish. 'Most of us don't know who our state reps are, except when we're going to vote, and we have to kind of figure it out. For those local offices that are our representatives, I think the ethnic affinity probably matters more.'

Again keeping it in the family, he quotes his sister.

'My sister Eileen is my bell-wether for how most of us process and think. She says, "When I get to a point where I don't know the candidates, I look first for the women and then for the Irish surnames." So for her, after Democrat comes women, and then Irish surnames.'

There are many ties between Baltimore and Ireland, outside of the Irish who fought in the American Civil War or built the railroad. O'Malley is wistful of where the connections lie now. 'It's really the history and music that's all that keeps us here, isn't it?'

And while it might not be an obvious spot for current immigration, it is ongoing.

'There's still newly arriving Irish people who come and make themselves part of the fabric of this city and its life. I mean look at Colm O'Comartun (one of O'Malley's closest advisers). He's from Dublin and was, in essence, the executive assistant secretary to the cabinet of the mayor of Baltimore and then performed a similarly high-level kind of secretary to the cabinet role when I was the governor of Maryland.'

O'Malley had stopped to exchange pleasantries with a man a little earlier on our walkabout of Baltimore. 'John Moore – who I shouted across the parking lot at – he arrived here from Monaghan and was involved in the reconstruction in West Baltimore of what became an Irish bar in an old, beautiful brownstone that was all fallen and deteriorated. Jimmy Fagan from Ireland opened the James Joyce bar here.' Another Baltimore Irish bar that regularly features appearances by the governor's Celtic rock band.

Martin O'Malley thinks American politics would not be the same were it not for the Irish Americans who have been involved since the very foundation of the United States.

'I think that the Irish are at their best when in politics in the United States, with the realisation that just a short time before, they were the oppressed in their own country. They were refugees, dying on the mid-Atlantic passage. Every people's story is different and unique, and you never want to fall into the trap of diminishing the suffering of another people by comparing it to what your own went through. But I think the Irish are at their best in American politics when they carry into that public arena an empathy and respect for what others have gone through. Especially for what others are still going through.'

Growing up in his house, like in so many others, pictures of the Kennedys graced the walls, to provide inspiration about what an immigrant family could achieve.

'The belief that one person can make a difference, that politics is a noble calling and is a high form of service to others. Sure, I mean, when I threw myself as a volunteer into a presidential campaign in 1984, and when I decided to run myself as a candidate, it was with an awareness and a connection to all of those emotions, expectations – and expectations become behaviour. And in public service, when I was at my best, it was because of empathy, understanding and appreciation for the suffering that other people are going through. And also that well-placed Irish disdain and hatred of injustice and bigotry.'

As a former Democratic presidential hopeful, governor, mayor and chair of the Democratic Governors' Association, it is not surprising that he has no good words to say about the current US President, Donald Trump, describing him as a fascist.

'What is all the more painful is that there were so many people with Irish surnames around a man who gives cover to white nationalism, white racism and people in cages on our southern border. It's appalling and I'm sure many of our ancestors are rolling over in their graves to see names like Conway and Kelly and Bannon associated with this horrible mistake we've made in the highest office of our land. But all presidents are temporary.'

While Martin O'Malley says he was ashamed to see so many Irish surnames in the Trump administration, he was not surprised. 'There's an expression that a penny always looks down on a half penny,' he says, somewhat philosophically. 'There's an ugly side to the American experience, the American dream and its realisation. And that it is how readily people like to dump on those that are trying to follow up the rungs of the ladder to success. It would seem to me that one of the pitfalls of affluence and "making it" is that as it becomes easier to get one's sons and daughters into previously exclusive Protestant colleges, it also becomes easier to fall into the cultural bigotries and the racism that we still have yet to exorcise from the soul of America.'

Martin O'Malley says his days of running for elected office are over. He's dedicating himself to campaigning for others he believes in, putting his mailing list to use across the country. In the Democratic presidential primary this time around, he will play no part, and early in the process decided to back fellow Irish American, one-time congressman from Texas, Beto O'Rourke.

However, as we walk around Baltimore and through City Hall, he is stopped frequently by well-wishers. He says he misses the limelight; the ability to be able to do things for people and to make a difference – and, as he puts it, to be useful. But in many neighbourhoods in Baltimore, he points to housing projects, churches, museums, sculptures and public projects that he initiated or protected. His portrait hangs in City Hall alongside the city's other past mayors.

The power and influence of this Irish American is built into the city in a timeless way. One public building project was a renovation of a fountain in front of City Hall. There is a decorative design built into the paving – in just one corner, there is a little decorative shamrock. O'Malley explains that the sneaky shamrock is something that he and his adviser, Dublin-born Colm O'Comartun, asked the designer of the fountain to put in. 'It's almost like an engraver's secret little sign: the Irish were here.'

THE OPERATIVES

AN IRISH PARTY

Madison Square Garden, New York City, 30 August 2004. It was the morning of the first day of the Republican National Convention. President George W. Bush was seeking a second term in office, and while he ran virtually unopposed to be the Republican Party candidate, he was expected to face stiff competition from the Democratic Party. Their candidate on that occasion was John Kerry, then a US senator from Massachusetts. The contest was tight and ultimately Bush won a second term in office, with 286 electoral college votes to Kerry's 251. The decision came down to the votes of a single state – Ohio.

Fought against the backdrop of the Iraq War and the September 11 attacks, plus the Democrats' angst over how the 2000 election had been declared, the road to the White House was always going to be a tense one.

Neither candidate had notable Irish heritage, yet the Republican Party had noted how the Democrats had mobilised Irish America to President Bill Clinton's benefit in the 1992 and 1996 elections and they were determined not to leave any votes behind.

Susan Ann Davis remembers the morning of 30 August clearly. She had nerves like she had never felt before. Resplendent in a blue buttoned-up jacket and with a bouncy blow-dry, she walked to the stage in Madison Square Garden for what would be one of the biggest moments of her political life. She was going to address the delegates of the Republican National Convention and the multi-million television audience and pledge the support of Irish Americans to President George W. Bush. It was the first time anyone had spoken at a Republican convention on behalf of Irish American Republicans.

'I'll never forget that moment. It was such a big deal to be a speaker, and for a woman, of course, it was a really big deal. All the guys that I'd worked with over the years, they formed this, like, gauntlet, a double line, and cheered me on as I was going up to the stage. It gave me lots of confidence, which I needed, because I was petrified!'

Her speech was short and praised the work that President Bush had done for American–Irish relations during his first term, for the peace process in Northern Ireland and for business links.

She ended with this sentence to applause: 'The ties between Ireland and America are strong and vibrant. From gratitude from millions of Irish Americans we declare our support for the re-election of President George W. Bush.'

Susan Davis also hosted the first Irish American party at a Republican National Convention in 2000 – something that had been a long tradition at Democratic conventions. In fact, it was seeing how the Irish American Democrats had organised that spurred her on.

'Republicans really started to engage, to be honest, when I started to engage them in a visible way. It's really terrific for Ireland because of the Irish–American relationship, in terms of the interest from the diaspora and supporting issues in Ireland.'

The Republican Party has had continuous engagement with the Irish American community since 2004, but Susan says it's only during presidential campaigns.

Susan Davis became president of the Irish American Republicans in the year 2000 and is now the chairperson of the group. She's been heavily involved for the last twenty years and says it takes up a lot of time. There just aren't enough people to engage the community for every single campaign in every single district, so the focus has to be on presidential elections.

She says most of the effort is from the grassroots up, with a seal of approval from the party headquarters, rather than the headquarters pushing a hard agenda. She says the Republican National Committee (RNC), which leads the party, will reach out to all sorts of people when a presidential campaign comes around, be they Italian, Jewish, Irish or any grouping. But the needs are different depending on each candidate. It can be pulling people together to go canvassing, raise some money or even do polling.

'It's hard,' she says, 'because you're just using your personal networks. It's hard when you don't have a candidate with any Irish heritage.'

Nevertheless, there was an Irish American outreach programme during the 2016 election, but it was a special interest group rather than being based around any particular Irish issues or any link to Ireland. So the group of Irish Americans for Trump did small fundraising efforts and canvassed their own contacts. This was mostly under the radar and on a local level, such as breakfasts for businesspeople. Susan thinks candidates don't pay much attention to small special interest groups, but of course they are grateful for any 'get out the vote' efforts.

When candidates don't have an immediate connection to Ireland, Davis says, Irish American voters will give party operatives like her a hearing on the candidate, but it is strictly as a general voter and not because they intend to be swayed. She cites the example of her campaign for Tommy Thompson, who was elected governor in Wisconsin. 'He's Irish American, but he's also German, so as a party we had an outreach to both ethnic groups. I would canvass and say, "I'm Irish, you're Irish, listen to what I have to say." But it had nothing to do with Ireland.'

She is crystal clear on whether she thinks there is an Irish voting bloc in the United States. 'Absolutely not. Not at any level. Not at any district. Not anywhere. It's a nice myth.' She thinks there might just about still be a Jewish voting bloc, but that's it.

Rallying Irish Americans to the Republican Party ahead of presidential elections, for her, is a way of reaching voters any way you can with everything you can.

'I'm an Irish American, but none of the issues for the US have anything to do with Ireland. I'm interested in education and financial issues and foreign policy and all of that – that's how you're going to get me. You're not going to get me on an Irish American question because there's just no connectivity. And now, with the younger generations, that diaspora is getting smaller and smaller.'

Davis' reasons for connecting Irish American Republicans are not solely for electoral engagement and to further some greater cause, like the peace process. Her views are in contrast to those expressed by many others, who are involved for exactly those reasons. For her, it's about belonging to a network, seeking a common thread with other people, particularly in the political world. Relationships and networking count for a lot in Washington DC and if a shared bond with Ireland offers an opportunity, it's worth taking. However, it's an opportunity that fewer are interested in.

She points to diminishing numbers of people who self-identify as Irish Americans. In the early eighties, when she first got involved, there were nearly 40 million Americans of Irish descent, but the number has dropped since then. In the 2017 US census, 32.6 million people, or 10.1 per cent of the American population, ticked the box for Irish ancestry. It's still a sizeable percentage of the population, but it's not only Irish heritage that people claim: they may also tick the box for German, Dutch or Hispanic.

Ultimately, though, Susan Davis says it's about the economy. 'People vote with their pocketbook first. It has nothing to do with your ethnicity. It's great during a campaign to rally people around – it's really great for Irish–American relationships in the diaspora because you start reminding people that they're Irish. And they're networking with each other and coming together for a cause, but actually, during the campaign, it's not for Ireland. It's for Trump or it's for Biden – it's for whoever it happens to be.'

Like many others, Susan Davis feels the Republican Party itself is shrinking, but she feels the Irish American interest in it is not. She argues that many more senior businesspeople are getting involved with the Irish American Republicans than with the Irish American Democrats.

'In the early part of this century, say 2002 to 2005, a majority of the Fortune 100 CEOs were Irish American, and they were Republican, and that was a big surprise to me. I actually commissioned research on it.'

The Center for American Entrepreneurship (CAE) has analysed the Fortune 500 data for 2017 and found that 43 per cent of those companies were founded or co-founded by an immigrant, or the child of an immigrant. It did the same survey in 2011 and that figure was slightly lower, at 40 per cent. Looking just at the top 35 firms, 57 per cent were founded by an immigrant or a child of an immigrant. Of course, they are not all Irish American – about 5 per cent were either Irish immigrants or the child of an Irish immigrant. Those companies include some of the greatest in the US – both financially and in terms of the role they play in American life, such as Ford, McDonald's, Monsanto, Avon and the parcel delivery company UPS, to name just a few. These figures demonstrate both the recent decline in Irish immigrants and the wider impact immigration is having on American life.

In the 2001 election, however, Susan Davis used the large numbers of Republican Irish American businesspeople as a bridge from President Clinton to President Bush. 'My whole motto was "Peace to prosperity". Bill Clinton had brought peace and now we had Bush and he was bringing prosperity. So it was a way to get people involved.' She started an Irish breakfast club for businesspeople. 'They were really interested in being with one another and in what was going on in Ireland from a business standpoint.'

Irish American Republican congressmen attended the breakfast club and she laments that there used to be many more than there are now. 'There's been a big change in that. Well, there's been a big change in the Republicans, period! The pendulum swings. But, to be honest, you had to be really persuasive to get people to care about Irish affairs. It was kind of not on the radar screen. If you're running a big corporation, you're focused on your own issues and your own needs.'

Although a loyal Republican, Susan Davis is not a huge supporter of Donald Trump, but she is happy to have a Republican president. While she may not agree with him on everything, she understands how Irish Americans were attracted to him and why Democrats in traditional blue-collar districts noticed that some of their loyal constituents have switched allegiances.

'Ronald Reagan was fabulous at pulling in working-class people that were Democrats and then Trump, I think, was just really successful in going after people who were disaffected, in any number of ways, including all of those who were Irish Americans. A lot of Democrats, certainly in this last election, just changed their stripes and went for the Republican side.'

But she cautions that it may not be a permanent switch. 'They just got ticked off at Democrats and that's probably going to happen again the other way around. Or we're going to find more and more people becoming independent.'

But ultimately, she thinks, Irish immigrants, Irish Americans, are going the way of a lot of immigrant communities who assimilate into the adopted society and identify more with just being 'a regular voter' than an 'Irish American voter'.

'Beginning with Reagan and continuing on, as Irish Americans became more and more successful, they began to move from being Democrat to Republican because they were way more interested in the economy and financial issues and all those types of things.'

Susan describes herself as a 'cheerleader' for Irish America, but she is pragmatic about the future. While she thinks it's great to see people with Irish surnames and Irish heritage in the White House and in Congress – for either party – she is worried that the level of influence is declining. She doesn't think it will disappear entirely, such is the reach of the tribe and the emotional connection that people feel, but the question is whether future generations will choose to act on it.

In 2019, three sitting conservative US Supreme Court justices claim Irish heritage: Chief Justice John Roberts and the two Trump appointees – Justice Neil Gorsuch and Justice Brett Kavanaugh. Justice Roberts is connected to Limerick and has a holiday home there. Justice Gorsuch's maternal line can be traced back to Donegal, and his family marks a great achievement for an immigrant family – his mother was the first female head of the Environmental Protection Agency. Brett Kavanaugh's great-grandfather came from Roscommon, and his mother was also a circuit court judge.

But, as Susan Davis points out, 'I certainly don't think a Supreme Court justice is making any decisions based on being Irish American.' It is on some level remarkable that being an Irish Catholic did not arise as a point of discussion for the confirmation hearings of Gorsuch[4] or Kavanaugh, when it was considered a significant risk factor in public life for a predecessor, Justice William Brennan (whose parents came from Roscommon), and also for President John F. Kennedy. For both of those men, critics were concerned that they would put loyalty to the Pope ahead of fealty to the country. In the televised hearings that Kavanaugh faced in

---

4 Gorsuch himself is Episcopalian, but his mother is Irish Catholic and he was raised in that tradition.

Autumn 2018, more focus was put on his teenage escapades and a historic allegation of sexual assault than on his religion. However, frequent references to GAA and trips to Ireland did pop up in the diaries Justice Kavanaugh kept as a young man.

Davis points to the efforts she makes personally to keep Irish culture alive in her family into the future. Like many Irish Americans, although she feels a primal connection to Ireland, she doesn't have a lot of information on her ancestors. She knows 'for sure' that there was a Frank Barry from Cork who came to the US in 1848 and settled in Wisconsin, in a place called Wilton, which had a big Irish community. It's believed he was related to John Barry, born in Wexford in 1745, who went on to become commodore of the US Navy and is known as its founding father. Her last name – Davis – is Welsh Irish, and she said her grandfather used to tell a story that they were related to Thomas Davis but she doesn't know for sure. But she's trying hard to keep the family traditions going.

'My nieces and nephews, who are in their twenties, I really had to engage them because their parents might not have. I always throw a St Patrick's Day party for my family so we can talk about culture and things like that. But if I didn't do that, it wouldn't have happened in my family – and I'm one of five. Now, of course, they all love it – I've done it for so long. I took them all to Ireland – otherwise they probably wouldn't have gone. But as a result, one of my nieces graduated from Trinity College and, again, none of that would've happened, even though her mother is as Irish American as the day is long. All of my siblings, with the exception of one, married people of Irish descent in one way or the other, but you need somebody in each family to keep it going.'

—

As with many things in the Irish American community, community and culture come first, and partisan politics come second. Susan Davis, as chair of the Irish American Republicans, works closely with Stella O'Leary, chair of the Irish American Democrats. As the countdown to one of many Brexit deadlines loomed large in March 2019, Arlene Foster, leader of the DUP and former first minister of the Northern Ireland Parliament before it collapsed in January 2017, came to visit Washington DC. Susan Davis and Stella O'Leary worked together to host a bipartisan meeting for Irish American businesswomen and female politicians.

Whereas Susan's ancestors came to the US sometime in the 1800s, Stella was born in Dublin, moved to the US for work, married an Irish American and never went home. She had no intention of getting involved in US politics but became sharply aware of the Irish American connections when, one year after she arrived in the US, the 'Irish' President John F. Kennedy was shot.

'My husband was from Boston, and we were already engaged at that point – we married here in Dublin in 1964. It was very traumatic, very intense. All his family came down to Washington for the funeral. They were broken-hearted. I watched his funeral from a window at 1000 Connecticut Avenue, up high, as they went by with de Gaulle and all the leaders, walking up to St Matthew's Cathedral, and Jackie in the veil and so forth. We were up there looking down.'

Surprisingly, for anyone who knows Stella O'Leary as the doyenne of the Irish American Democrats organisation, she had nothing further to do with American politics for 30 years.

Then, in 1996, President Bill Clinton was facing re-election, and although he had been heavily involved in the Irish peace process, his work was not complete. A group of well-connected Irish Americans met in Washington DC to see how they could help him get re-elected. She was appointed the leader of the group and admits she knew absolutely nothing about US politics or how to set about fundraising.

'I called the Federal Elections Commission the next day and said, "What do I do?" And they said, "You just come down here, fill out a form that you're starting a political action committee and then it's up to you – go raise money." So I did that, and we started doing posters and getting in touch with Irish groups. Then I found that there was support everywhere. You could talk to Boston, Cleveland, San Francisco, Omaha, Pittsburgh, Philadelphia, anywhere. It's still the same today. They would rally right away for their candidate – in this case, Clinton.'

And so President Bill Clinton was re-elected.

John McCarthy is one of the younger American political operatives dedicated to furthering Irish affairs. He is a member of the Irish American Democrats group, chief of staff to the only US congressman with an Irish-born parent, Brendan Boyle, and headed up the Irish outreach programme in the Democratic National Committee, the party's headquarters. He is fiercely proud to be Irish American, but even he admits his ancestry is fuzzy.

'My family came in here in the 1700s and that's about all we know. My parents are very proudly Irish American. We have shamrock things all over our house and St Patrick's Day was always a very big deal, but my parents have never been to Ireland. It is totally Irish American, not tied to the island. There's such an American cultural pride to being Irish and it means something somewhat devoid of the island. It's an "Americana" experience and it conjures up notions of family and faith. I just think it is too ingrained here to ever go away.'

His family went to the local Irish festival every year. On St Patrick's Day they all wore McCarthy T-shirts. His father was in the Ancient Order of Hibernians; his uncle was the local head of the chapter. His family are exactly the sort of people who were mobilised to vote for Bill Clinton.

McCarthy says Clinton understood the need to build broad coalitions, and he believes the 'Irish American community' does not necessarily mean all 33 million people who identify as Irish American by ticking a box on a census form.

'The Irish American community are the people who are actively engaged around the issue of Ireland or Irish America, and they were thrilled with President Clinton. I think Bill Clinton saw that and he himself was the one who could have helped move that up in the Hillary Clinton campaign.'

And so why was it not possible to copy and repeat this for the 2016 candidate, Clinton's wife, Hillary?

'I think Hillary was really well served by having John Podesta [chair of Hillary Clinton's 2016 election campaign] there, because John Podesta was a huge advocate for this type of work. He understood that these communities are valuable and meaningful and can be engaged. I think a lot of the younger staffers who came from more coastal cities didn't grow up around this cultural identity and sense of community. But where I'm from, in the middle of nowhere New Jersey, there's a big Irish high school, there's a big Italian high school. Their families were around this and that's what they talked about.'

That 'Irish' high school was called Red Bank Catholic, the mascot was a shamrock and the nickname was the Caseys. His hometown, which he describes as 'middle of nowhere', is Keansburg, and the coastal area it's in is known in New Jersey as the Irish Riviera because there are a number of little shore towns mostly filled with older Irish immigrant families.

He says President Clinton and Podesta pushed that type of retail politics, targeting smaller local groups and individuals personally, but it didn't wash with others running Hillary for America.

'I think a lot of the folks at the more junior levels of the campaign didn't see the value of it, so it didn't become a cultural imbed.' He thinks that reluctance to engage in grassroots political outreach is a problem at the highest level in the party organisation too.

Many in the Irish American community feel they tipped the scales in Bill Clinton's favour in his 1996 re-election campaign, but Stella O'Leary goes even further – she believes the Irish lobby actually saved his career.

'The support from the Irish made a huge impression on US public opinion, that the Irish were so heavily behind him. Otherwise, he might've gotten impeached by the Senate. He was impeached by the House, as you know. The pressure from the Irish played a huge role in preventing the Senate from impeaching him.'

But things have changed in the US and in Ireland. Stella believes the influence Ireland enjoys has not necessarily lessened, but it has changed.

'Considering the size of the island and comparing the influence of Ireland to the influence of virtually any other ethnic group I can think of, it's way, way greater. There is nobody on Capitol Hill, no congressional member, who'll close their door to an Irish request – none. They've no reservations about it. They don't ask questions. They say come on in and tell me what you have on your mind, and so forth. Whereas, they would be avoiding some groups. So it's unusual; it's exceptional. And, of course, that's not just politically: that's everywhere. My son has said to me, "'The best thing you ever did for me was give me the name O'Leary."

When it comes to a voting bloc and an ability to sway an election, Stella O'Leary doesn't think there is a national bloc, but some clusters of Irish remain in some localised areas. Sensing that Hillary Clinton needed all the help she could get in certain areas in 2016, O'Leary, a friend of the Clintons, directed some of the firepower of the Irish American Democrats organisation to Ohio.

'For the last election, I opened an office in Cleveland, Ohio, in Cuyahoga County and I picked an all-Irish district. When you took a look at the rolls in that area, we had about five thousand families. We opened an Irish office – posters and vote Hillary and all the rest. Then I published a pamphlet on what she had done for Ireland and we distributed that to all the households in that area. She lost Ohio; she won that area. Now, I'm not saying it was all our doing, but she won Cuyahoga County [by] as much as she lost the other by. So if you can touch these people and say Ireland needs that person, they vote for that person. But you don't have them all in the one place like we had there any more. And Omaha, Nebraska, we have the same thing. We have a whole group of Irish there.'

Jim Cavanaugh is a county board member in Douglas, Omaha, Nebraska, similar to an Irish county councillor. He proudly boasts that Omaha is one of the top ten American cities for people of Irish descent. However, just as the ethnic Irish groupings were not always welcoming to newer immigrants in other parts of the country, the same was true in Omaha. For example, on one occasion, angry at the murder of Irish police officer Edward Lowry by a Greek native, an angry mob of reportedly[5] up to 3,000 Irish essentially ran the Greek immigrants right out of South Omaha (then a separate city). The riots were racially motivated as, according to the *New York Times*, the crowd was riled up by local politicians – including City Attorney Henry Murphy, and they 'assaulted many Italians and Romanians, who were mistaken for Greeks'. Murphy, of immigrant stock himself, had stoked the crowd by describing what he referred to as the 'undesirability' of the Greek people, saying 'the blood of an American is on the hands of those Greeks … and some method should be adopted to avenge his death and rid the city of this class of persons.' This sentiment is not dissimilar to the anti-immigrant views expressed by other Irish Americans currently in positions of political and civic leadership.

The reason for the concentration of Irish heritage stretches back to the aftermath of the American Civil War, says Jim Cavanaugh. 'They completed the transcontinental railroad and that railroad

5 As reported in the *New York Times* on 22 February 1909. 'South Omaha Mob Wars on Greeks'. https://timesmachine.nytimes.com/timesmachine/1909/02/22/101816600.pdf

and the telegraph that ran across it was the single connection for some time between the east coast and the west coast of North America – all commerce, all communications. It would be like owning the only telephone between New York and San Francisco. We were the only railroad, so everything came through here.'

The railroad was primarily constructed by Irish workers through the middle of America (just like the Baltimore–Ohio railroad on the opposite side of the country), and by Chinese workers on the west coast, so that made Omaha something of a hub for Irish immigrants. Jim Cavanaugh says that's when his father's ancestors – who were so-called Famine Irish – came, but his mother's family came later as part of the 'Revolutionary Irish', those who fled during and after the Rising, the Civil War and the War of Independence. He somewhat conspiratorially adds that his grandmother was an undocumented immigrant when she entered the US and lived there illegally for twenty or thirty years before she became a US citizen. But she never went back to Ireland, never again saw a living person she knew from Ireland, and he says that 'tore the heart out of her'.

'The turning point of President Kennedy's presidency was the breaking of that glass ceiling – not only in government but all across the socioeconomic sphere. It's now unremarkable to see an Irish American in any capacity at the top of our society, whereas there were huge debates when President Kennedy was elected about whether or not we could allow an Irish American serve as president.'

But preserving the influence and the culture of Irish America into the next generation takes effort. 'We have gone out of the way with the next generation in our family to make sure they're well aware of their Irish heritage and roots. Children of mine have been educated in Ireland, and we go to Ireland regularly and spend time there.' He's also secured grants from the Irish government to bring Irish art and artists to Omaha, and is involved in the sister city programme – Omaha is twinned with Naas, County Kildare.

Nebraska is a Republican state, although the area around Omaha votes Democrat. The Irish Americans throughout the state are changing their colours.

'It's probably a by-product of this affluence that has occurred over generations. They used to be almost exclusively Irish Catholic Democrats and are still overwhelmingly Catholic. The difference is they're not all Democrats any more. What are called Reagan Democrats were based on Irish American Democrats who became Irish American Republicans around the time of Ronald Reagan's ascendancy. I think part of that is the Republicans are much more popular with wealthy people than the Democrats are. As the Irish rose in the socioeconomic sphere, the children of that experience had a tendency to listen more to Republican economic theory,' says Cavanaugh.

Mentioning Hillary Clinton's loss in 2016 is a little easier for Stella O'Leary now. The passage of time is a great healer and the focus for her and other Irish American Democrats is now on the 2020 race. But she was one of those convinced of a Clinton victory in advance, and someone who may have been considered a potential US ambassador to Ireland. In fact, so many Irish American Democratic political operatives had helped the Clintons out over the years that the running Washington DC dinner-party joke was that at any Irish event in the US in 2016, there were probably five to seven people in the room who considered themselves to be 'the next US ambassador to Ireland'.

Given that so many people felt they were crucial to the election of Bill Clinton in 1992 and 1996, why didn't the same effort mobilise for Hillary Clinton in 2016? 'We just don't have any dramatic issues. With Hillary's election campaign, we were saying "save the peace process" but the peace process wasn't in danger. If we were running an election today, we'd put out leaflets saying the British are going to put a border back in Ireland. That would influence some voters.'

Although the Irish American Democrats organisation is registered as a public action committee, or PAC in American political speak, so it can fundraise effectively, its efforts are not coordinated by the headquarters of the Democratic Party, the DNC.

The DNC has the National Democratic Ethnic Coordinating Council (NDECC), which is not exclusively for Irish/Irish American voters. It's an official allied organisation of the party, similar to the College Democrats of America or the Women's Leadership Forum, which are the other two allied groups. It gets two seats appointed to the Democratic National Committee.

John McCarthy, the Irish American from New Jersey, now working as chief of staff for Congressman Brendan Boyle on Capitol Hill, is also the ethnic coordinator for the DNC, a role he had during the 2016 presidential election, along with being the faith and heritage outreach coordinator for the Hillary Clinton campaign. Since becoming a congressman's chief of staff, the DNC ethnic coordinator role should move to someone else, but as that hasn't happened yet, he is still running the outreach side of things.

As there aren't currently any appointed staff for the NDECC, McCarthy thinks that's not where the party's intentions lie at the moment. He points out that there are currently no staff for rallying unions or business groups or similar coalitions. There were, until 2015, and then they were absorbed into the wider Hillary for America action group and have never been replaced within the party structure.

But, he said, the DNC knows that come 2020, in particular, it will be all hands on deck, so he expects to get called upon again. 'We're going back on the road doing a listening tour. We'll do a series of meetings in all of the swing states, just to engage across the different ethnic voting blocs.' And that's across all ethnicities, not just the Irish.

'The whole purpose is to organise now so that whatever apparatus we can build up can be handed over to the nominee, and then when somebody becomes the nominee, they also get to change all the staff at the DNC as well.'

Stella O'Leary thinks that the DNC could be doing more to use the Irish Americans. She says that anything she and the group organises, they tend to do themselves. The DNC will put out 'buttons and bumper stickers' for most groups, and that's about getting the voters out to the polling stations.

'They don't use the ethnics as well as they could. There are definitely places like Michigan where there are ethnic groups who would be moved if, say, the person running for vice president came and spoke to them about their native country, but of course somebody may also be voting only on education or taxes or something.'

But John McCarthy thinks that the DNC's lack of perceived interest in lobbying Irish Americans has less to do with where Irish affairs and the Irish community are at currently and more to do with the wider US voting picture. He thinks there's an 'identity struggle' within the party, and it breaks down into two schools of thought.

'On one hand, there's the old LBJ model of having everybody under one big tent. And then there's the other part that says we no longer need these types of voters – all we need to do is run up the margins on our base, increase the numbers of our most typical voters. And if a candidate becomes a nominee under that philosophy, this kind of outreach won't happen because Irish Americans under that philosophy are not part of the team.'

But having said that, he does still feel the Irish Americans matter – perhaps not to get the party's nomination to run for the White House, but it would be difficult to win the presidency without having anyone with an Irish identity vote for the candidate.

'It's not that you need the Irish American voter alone – you need the Polish, Italian, you know. You need the working-class voters in those old ethnic enclaves outside of Cleveland and Columbus and Philadelphia.'

McCarthy thinks it's got a lot to do with how identity politics or local politics or voter organisation is approached. It's essentially a style choice at a time when the US political system is going through a period of major change.

'I think that some folks in the party would sit and look at this kind of work as silly or outdated – and to some extent I think the folks who are interested in this kind of work do need to reinvent themselves. The Irish vote doesn't exist the way that it did in New York in the early 1900s or the late 1800s, but for people who care about politics and the Democratic Party, that old model still applies to new immigrant communities.

'Whether it's the new Bosnians in Iowa or the new Arab American communities outside of northern Virginia, the lessons we've learned from how the Irish came to America still apply to these communities. They're highly networked, they know each other, they have their own ethnic media outlets, a lot of them revolve around some type of religious service and those kinds of communal hubs, so there are lessons to learn out of that.'

John McCarthy also does not think there is an Irish voting bloc in the US. 'The best way to think of it is like this. I don't think there's an Irish vote – there's an Irish constituency, which is different. When you're working inside the Democratic National Committee or a campaign, there are all sorts of constituencies that you think, if people are willing to identify that way, you need to find ways to engage them under a hat they feel comfortable with. So there are clear voting blocs for Latinos, African Americans, women, and you can kind of judge to a certain extent and say they would all vote following this pattern – Irish people aren't that way.'

He would urge moving away from the term 'voting bloc' – it's more of an Irish lobby, he feels, 'and it's even more than that. It's an advocacy arm.'

He believes there are many reasons why Irish Americans and Irish American politicians can have the level of influence they do, and why the Irish enjoy the level of access that they do. 'First reason is the numbers, of course. Number two is "brand Ireland". It's incredibly strong, stronger than ever, it seems. I see it all the time when I invite people to Irish American events, even if they're not Irish – it conjures up something that's going to be a good time; people will be lively. It conjures up an image that people want to be a part of.' And that impression is something he thinks has been the result of a well-cultivated effort by successive Irish governments and by Irish America itself. An effort that has largely paid off, if one gives credence to the financial value afforded to the concept. Marketing and finance company Brand Finance attempts to quantify the value of nation brands each year. The latest figures available (for 2018) show the value of 'Brand Ireland' to be worth $469 billion, a rise of 26 per cent on the previous year.

John McCarthy's family moved over in the 1700s. Although his parents have never been to Ireland, and have no interest in tracing the minutiae of their heritage, they are 'very proudly Irish' and he had a classic Irish American upbringing with festivals and shamrocks and music and stories.

'From the Irish America standpoint, it was the cultural club to be a part of. It gave people something to do on weekends; it became part of people's social fabric. Again, totally devoid of political activism, but it was something I always really cared about.'

He thinks these social hubs, like Irish festivals, St Patrick's Day parade committees, the Ancient Order of Hibernians and so on, are what has allowed the Irish to be so politically organised and to be so influential.

'When people decide to run for election, there's an organised community already waiting. When you conjure up that image of Irish America – charming, funny, interesting – I think people are drawn to that. That works in politics, too. I think Irish Americans are naturally inclined towards politics. I think the Irish are unique in that, as immigrants when they came in, they instantly got involved in politics, whereas a lot of new communities, when they come here, need to be engaged.'

But McCarthy argues that Irish America, to some extent, has nothing to do with the island of Ireland. He often doesn't recognise the Ireland that he hears some Irish Americans talking about. 'There's actually a danger when some of that Irish American pride tries to translate into political activism, because I think if you're going to advocate effectively, you need to have a clear picture of what modern Ireland is and what a mutually beneficial relationship is.'

Many are stuck with an image of an impoverished, famine-ridden country that their great-great-grandparents fled, and they have grown up listening to those stories, not stories of prosperity. 'Ireland is an amazing first world economy now. When we have this notion of small villages and thatched roofs, that's not the Ireland of today, and when that is referenced on a global stage, it actually puts Ireland in a weaker posture.'

There is also the charge that Irish America is far more conservative than Ireland is. 'Irish Americans still have pockets of conservatism around the country, and for them, they look at a country where marriage equality wasn't allowed a few years ago and now there's a gay taoiseach, right? The country is progressing in incredible ways that Irish America might not have caught up to yet.'

He argues that, although many in the Irish American community want to feel included in the fun, they also want to be involved with Irish issues – that is, as long as there are issues. 'It's hard to keep people engaged when there is nothing going on, there is nothing to lobby for, there's nothing to stay politically relevant on.'

With the uncertainty and risks around the peace process and a return to a hard border created by Brexit, there is a new issue for the community to get involved with. 'Brexit, I think, re-engaged a lot of people. The tough thing is, here in the United States, at least, it came after the curve, right? No one thought Brexit was going to happen, but it did. Then suddenly people said, oh, I've ignored this for almost twenty years, I need to get up to speed about everything that's been happening in Ireland. I think a lot of people came into it with a point of reference that was still in the late nineties, and a huge and successful economy, a diverse and talented workforce, a modern Ireland, happened in those twenty years. I think Irish America had to wake up and realise that really quickly.'

While Stella O'Leary feels many ethnic groups look at the Irish as a good example of an organised diaspora, she feels the Jewish community do it even better. The level of influence they can wield is superior. When she was setting up the Irish American Democrats PAC, she met with her Jewish counterpart at AIPAC (American Israel Public Affairs Committee).

'I got the best advice I've ever gotten in my life. He said, "What is your mission? What do you want?" I said, "Well, I want to see sustained peace, justice and prosperity in Ireland." He said, "Don't let anybody ever discuss anything else with you except that. And don't answer any questions about anything else." The first 40 phone calls I got were about abortion. If I hadn't had that advice, I would have been searching for what to do, and how to do it. So, I said I'm not an expert on it. I know nothing about abortion. It was great advice. You know how small an ethnic group the Jewish are? They're, like, 9 per cent. The Irish are about 10 per cent of the population. They had several political action committees. The Irish had none.'

Presidents and vice presidents frequently address the AIPAC's annual get-together, and although it is a public affairs committee rather than a political action committee, it does not engage in fundraising for campaign financing directly itself. However, a study by *The Guardian*[6] of campaign finance information relating to the 2018 midterm elections showed that pro-Israel lobby groups contributed a total of $22 million to campaigns.

Stella O'Leary maintains that Irish interests don't need massive fundraising and powerful lobby bodies because there are so many leaders in Congress and in successive administrations who will fight for Ireland's cause, without needing huge campaign donations, because they have an emotional, tribal, ancestral connection to the country.

She agrees with the viewpoint expressed by Susan Davis of the Irish American Republicans about the swing that's taking place with those who claim Irish heritage – from the left to the right, from blue to red.

'I think that whole phenomenon is startling. I don't have any complete explanation for it. It never made any sense to me because many of the ones I know were not very far from Ireland. They were the children of immigrants and they moved out to a suburb and became Republicans, and then some became intolerant right-wing Republicans. I can only say, based on history and having studied and read a great deal about what happened in the South with the Ulster Scots and so forth, that people used racism as a tool to keep the blacks down and the whites up. It suits their purposes, or it did suit their purposes historically, because they could pay the blacks low wages, deny them rights. Because there was a lot of racism in the Irish, those early ones, and more recently. I've heard them. For example, when I went to the South first I was introduced to a woman that became a great friend of mine. Her mother was from Cork and was alive and well. I went to the Martin Luther King demonstration down on the National Mall. I took the bus from South Dakota Avenue downtown and she said to me, this woman,

6 Perkins, Tom 'Pro-Israel donors spent over $22m on lobbying and contributions in 2018', 15 February 2019, *The Guardian*. <https://www.theguardian.com/us-news/2019/feb/15/pro-israel-donors-spent-over-22m-on-lobbying-and-contributions-in-2018>

daughter of a Cork-born woman, said to me, "If one of those n—hurts you, I'll make sure there's revenge." So that's straight out of Ireland. I was horrified.' She sighs, pauses and continues. 'I've seen racism really rampant in some Irish American Republicans.'

Of course, Stella O'Leary, as a staunch Democrat, is pointing to racism in Republicans, but it cannot be categorically stated that all racists are or were members of the GOP and no Democrat has racist tendencies. For just as many Irish immigrants and the generation that immediately followed them were driven by hard work, public service, a commitment to social justice and perhaps an activist motivation, some were not inspired by any of that and were less savoury characters.

As far back as President Abraham Lincoln's Emancipation Proclamation in January 1863, Irish immigrant members of the Democratic Party had expressed their dissatisfaction. From the time of Lincoln's election in 1860, the Democratic Party issued warnings to its Irish (and German) members about the expectation that the abolition of slavery would lead to labour competition when freed southern African Americans would supposedly head north and challenge the Irish and German immigrants for low-paid jobs.

Furthermore, a change to the draft laws in March 1863 meant that all male citizens between 20 and 35 and unmarried men between 35 and 45 could be drafted. Black men, who were not considered citizens, were exempt from the draft. The first round of the draft lottery sparked the Civil War draft riots, during which time a mob, comprising many Irish, terrorised black men, women and children, including carrying out some lynchings.[7]

It would be naive to assume that all Irish immigrants who have come to America have made the country a better place, or have all been upstanding members of the community. There has

7 Harris, Leslie M. (2003). *In the Shadow of Slavery: African Americans in New York City, 1626–1863*. University of Chucago Press, IL

traditionally been a dark side to both political parties. The views held by some Irish Americans implementing the controversial immigration and border control policies of the Trump administration, and the voters who support them, have reminded critics of this.

—

Traditionally, many of the trade unions, particularly construction-based unions, were a good hunting ground for Irish American Democrats. The blue-collar, hardworking, politically engaged Irish workers were embraced by the party and in turn embraced the party. That changed twice in recent times. Many backed Ronald Reagan in 1980 and they backed Donald Trump in 2016. Stella O'Leary says she was surprised how the unions moved in the Trump election and says her friend, Leitrim-born Jim Boland, president of the International Union of Bricklayers and Allied Craftworkers, was 'furious'.

A union endorsement is worth a lot to a presidential candidate. It's worth a lot in money, but it's also worth a lot of votes. Boland says unions are heavily involved in Democratic politics, and while they don't always agree with Democrats, he says they're 'intrinsic' to the Democratic Party.

They usually have presidential forums in the run-up to an election and they invite all the Democratic candidates during a primary season. He says they 'rarely' invite Republican presidential candidates to talk to the unions' members, and in the 2015/2016 primary season, he didn't invite any Republicans.

Jim Boland says they decide on who to endorse based on who is going to be good for the members. He says some of the members, about 25 per cent, will listen to their employers more than the union when it comes to deciding who to vote for. In 2008 and 2012, about 70 per cent of his members voted Democrat, for Barack Obama.

The reason his friend Stella O'Leary says Jim Boland was 'furious' in 2016 is because they endorsed Hillary Clinton, but this time 44 per cent of his members voted for Donald Trump.

Jim Boland said he did a lot of soul-searching as to what the union leadership had 'done wrong'; what had caused so many of his members to break with the union party line.

'I think it's because he was able to present himself as a populist. He said he loved bricklayers on television. I've been a bricklayer all my life – I have never said "I love bricklayers" to an audience of ten million people. So you have to fight back against that, especially when he says, "I'm so good for you. I protect you from the elites. I'm going to do all of this stuff." He painted himself as somewhat different to what he is.'

Jim Boland is also chairperson of the immigration committee at the powerful American Federation of Labor and Congress of Industrial Organizations (AFL-CIO), the largest union group in the United States. Immigration is a huge point of concern for the construction workers he represents as so many are immigrants. He says they never ask members whether they are illegal or have documents to work – that is the employer's call, as far as he's concerned.

'I've been an immigrant all my life. Here I am. I've never been asked for anything except my name, date of birth and social security number, and I've been working in America since 1970. That's all I've ever been asked for, and I've probably worked for 70 employers. I say that around a bunch of Latinos, and they say, "Ha ha." I know Irish people have gotten deported too, but that's been my experience. There's raids and stuff all the time at plants where there's a lot of undocumented people. They ship them out. Obama did it too just as much as these guys. Hence traffic at the border is at a 40-year low of people trying to come in because it's not a very attractive business. So they don't really need a wall. They need immigration reform since there's such low unemployment and we're an ageing population. We need to bring in people to this country. The Europeans aren't coming here. We need workers. What we really need, and nobody wants to talk about it, since there are about eleven million immigrants working in this country who are not permanent citizens, is comprehensive immigration reform. It's ridiculous not to take it on, but there isn't the political appetite to sit down and actually solve that problem.'

—

Engaging Irish America on issues often has less to do with Republicans and Democrats and more to do with the betterment of Ireland. For this reason, many of those who work for either party also work together throughout the year. Republican and Democrat operatives each admit that the Democratic Party has been better at signing up Irish membership than the Republicans have.

Another Irish American, John Feehery, is an adviser for the Republican Party. Originally from a suburb of Chicago himself, he has great-great-grandfathers from Ireland on his mother's and his father's side. Like many Americans, he's taken a DNA ancestry test, which showed that he is 90 per cent Irish. Looking at what the Democrats have been doing locally, he says, he's been trying to convince the Republican Party hierarchy for years to have a more formalised outreach system to Irish Americans. 'The party doesn't have that legacy. We tried to build it. I've tried with different folks through the years at the RNC. It just doesn't catch. The Republicans aren't really good at playing those identity politics.'

Nevertheless, plenty of individual candidates do tout themselves as Irish. There may not be an Irish voting bloc, but it can be worth a few votes. Feehery says they regularly have candidates with an Irish name and so they make their campaign sign feature green.

'It still has an emotional value. There's a lot of people out there, if they don't know a candidate, they'll vote for the Irish person, the Irish name.' He thinks it can sometimes be worth two or three percentage points in an election – the difference potentially between winning and losing.

And although the issues to engage with aren't as serious as perhaps they once were, Feehery says Ireland has an important role to play in the United States when it comes to explaining the European Union, which, he says, can be 'hard to make heads or tails of' for American policymakers. This, he believes, is where Ireland can have a lot of leverage into the future.

'They can help navigate what's going on in the European Union, and that's a huge advantage for Ireland. Most policymakers here don't have time to figure out Brexit so having those Irish delegations come over here and explain some of those things is really important.'

The biggest challenge facing the Irish in preserving their level of access and influence, in his view, is the diminishing numbers of immigrants in recent decades because of restricted immigration flows. John Feehery says the Democratic Party is not necessarily looking out for the interests of Irish people on this, although the Irish may think it is. Appealing to as many immigrant communities as they do, he says, the Democrats have to please everyone and can't offer any true benefits to the Irish exclusively.

Modern-day Irish immigrants are not impoverished people who speak only Irish and cluster together. John Feehery describes them as 'really sophisticated', with a higher education, and they are coming, not out of financial hardship, but because 'they want to see the world, and do something different with their lives, and maybe just want to get out of Ireland'.

But it's harder than ever for them to come to the US and they have the choice to go anywhere – this, Feehery says, is something that many Republicans have been trying to impress upon the president. 'The biggest problem for the United States is, when you won't have really well-qualified immigrants from Ireland, it actually hurts America because you're losing out on talent. It's a completely different dynamic. When we don't allow the Irish to come into the United States, we're really not allowing the next job creators to come, as typically they get right into it and contribute immediately to the economy, so that's a huge problem from the United States' perspective.'

The Irish vote – such as it is – can at times be mixed in with a Catholic vote. In parts of Irish America, being 'Irish' is synonymous with being 'Irish Catholic' and the terms will be intermingled. A lot of Catholic voters are Irish but not all Irish voters are Catholic.

In the 2004 election, one of the biggest indicators of whether people voted for the Democratic nominee John Kerry or the incumbent Republican President George W. Bush was church attendance. The more likely you were to go to church services, the more likely you were to vote for President Bush, and vice versa.

In addition to his Irish American outreach for the Democratic Party, John McCarthy also played a role for the Hillary Clinton 2016 campaign in rallying Catholic voters. He says there are lots of people in the Irish American lobby who are Catholic, but he doesn't think they lead on any of the traditional Catholic issues. For them, he says, church attendance would not be an indicator of anything. There are distinctions between being Irish, Catholic and Irish Catholic but he says they are often all grouped in together – incorrectly, he argues.

John Feehery says you can essentially split the hairs even further – he says it's about tribes. His connection with his Irish American side was a quest for identity, to seek out a tribe that he felt he belonged to. He says now, 'I identify my tribe as a conservative Irish Catholic.' There is another tribe that comprises liberal Irish Catholics.

In Washington DC, where arguably the only religion is politics, he says there is a 'subset' of Irish Catholics, and the conservatives and liberals within the group 'look out for each other'.

'We have a decent relationship and we all go to the same church. We drink in the same pub, the liberals and the conservatives, because you have that in your own Irish family. You have your liberal cousins and your conservative cousins, so that's not that unusual!'

But still, so many US elections come down to where candidates stand on the key issue of abortion. Some voters will choose candidates based on whether they are pro-life or pro-choice, regardless of party, such is their own personal belief system. They want elected representatives, particularly at presidential level, to reflect their own values.

Crucially, though, John McCarthy says that voters want authenticity in candidates, and he gives a rare moment of praise to the current resident of 1600 Pennsylvania Avenue. 'Say what you will about Donald Trump, all the lies, all of the other statements, he is authentically himself. I do think, to some extent, people found that level of authenticity refreshing. He says what's on his mind. You might not agree with him, but at least you know where he falls on things. That's what people want.'

John Feehery says he wasn't at all surprised that so many Irish Americans voted for Donald Trump, and he thinks a big part of it was to do with religion and the abortion issue.

'Quite candidly, the Democrats have been pretty anti-Catholic, and I think it hurt them with a lot of our folks and a lot of the Irish Catholic Democrats,' he says. 'There's a lot of Irish Catholic Republicans who work on the Trump campaign, and they've worked on a message. Abortion is a big deal, to be honest with you. You can't really be pro-life and a Democrat these days.'

He says that religion, and in particular the view of the Catholic Church, is another difference between Ireland and Irish America.

'The Catholic Church is still a source of pride, warts and all, whereas in Ireland it was a monopoly that had to be broken – it had such a stranglehold on the consciousness of the Irish.'

Feehery says he is fascinated by that breakdown in the power of the Catholic Church in Ireland. Its place in Irish society has changed so fast, and has been so dramatic, he says. 'We don't have the same situation in the US. A lot of people are disaffected with the Church and are definitely pissed off at the child abuse scandals and a variety of other things, but I still think the Church is pretty strong here as a minority religion. It's not a majority religion.'

Politicians have often pursued the 'Catholic vote' in the way they also court the 'evangelical vote'. There are approximately 51 million Catholic adults in the United States, and if they all voted one way, they would constitute a massive voting bloc. They are credited with voting to get President John F. Kennedy elected, the first and only Catholic US president. While once they were more likely to vote Democrat, they have more recently been quite evenly split between both parties.

A preliminary analysis of the 2018 midterm elections shows that white evangelical voters continued to vote Republican in massive numbers – about 75 per cent of them backed Republican candidates. However, data from the Pew Research Centre shows that, for the first time since the 2006 midterms, a slim majority of Catholic voters backed Democrats over Republicans.[8] The change represented a swing of just 5 per cent, but Catholic voters are statistically more evenly spread between Democrats and Republicans than evangelicals are. Pew Research's study of religion in America carried out in 2014 found that American Catholics are overwhelmingly white or Hispanic. Black and Asian Americans

8 Elizabeth Podrebarac Sciupac and Gregory A. Smith, 'How religious groups voted in the midterm elections', 7 November 2018, www.pewresearch.org. <https://www.pewresearch.org/fact-tank/2018/11/07/how-religious-groups-voted-in-the-midterm-elections/>

only account for approximately 3 per cent of America's Catholics.[9] Repeated surveys have found that white Catholics tend to vote Republican and Hispanic Catholics tend to vote Democrat. This played out in the 2016 presidential election, where 60 per cent of white Catholics voted for Donald Trump and 67 per cent of Hispanic Catholics voted for Hillary Clinton. There is no recent information breaking down US Catholics by ancestry, so we can't know how Irish Catholics specifically cast their votes.

'A lot of Irish Catholics were going to vote for Trump on the abortion issue, but a lot of them were going to vote for Trump on a variety of other issues too because they just liked that he was tough, and if you're an Irish blue-collar worker who works in a factory, he was talking to you, and the fact that he's politically incorrect was, I think, very appealing to a lot of Irish Americans. I think it was no surprise that he got a lot of votes.'

Donald Trump scored well with white men in 2016 and he will be banking on this support base for his re-election in 2020. John Feehery thinks he will be safe there. 'There's a lot of Irish American Catholics who just don't like being PC, and they don't like this reflexively anti-white-male motif that the Democrats seem to be hammering on about. A lot of these are white males, and what they see is this assault on the traditional family, which is a big deal with a lot of Irish Catholics.'

As occurred in the 2016 campaign, Feehery says the issue of civil and family rights can alienate certain voters, such as the focus on transgender rights. 'The culture's really moved significantly to the left, and I think for a lot of Irish Catholics that's worrisome. My guess is that Hispanic Catholics will vote much more for the Democrats, as they have in the past, and Irish Catholic Democrats will vote for the Republicans – and vote for Trump again.'

9 Michael Lipka and Gregory A. Smith, 'Like Americans overall, U.S. Catholics are sharply divided by party', 24 January 2019, www.pewresearch.org <https://www.pewresearch.org/fact-tank/2019/01/24/like-americans-overall-u-s-catholics-are-sharply-divided-by-party/>

FAR AWAY IS SOMETIMES GREENER

There may not be a prouder Irish immigrant in Pennsylvania than Francis Boyle. He left Glencolmcille in County Donegal aged just nineteen, following in the footsteps of so many others before him to Philadelphia. As a young man attending various dance halls, he met the daughter of Irish immigrants from Sligo, who had moved to Philadelphia when they were in their twenties. He got a solid, paid and pensionable blue-collar job in the Sanitation Department of Philadelphia and he and his wife went on to to have two sons, Brendan and Kevin.

'I can't remember a time when I wasn't conscious of an Irish identity in our family,' says the elder son, Brendan. His mother's parents lived just a block away, so of the four family members he was surrounded by, three had native Irish accents. 'And we got weekly newspapers in the house like *Irish Echo* and *Irish Edition*. Both my brother and I always had a strong interest in Irish singers, Irish culture, Irish politics.'

But when it came to running for elected office in Philadelphia, being Irish, having an Irish family or whether there was any intrinsic political value in having a name like Brendan Boyle was not something the young political hopeful thought of. 'I'm mystified that anyone abroad would think that there's this great political payoff for being involved in Irish affairs. When I'm involved, like recently on the Brexit issue, it's not because there's any great political benefit back home. Most of my constituents, the overwhelming majority, are not of Irish descent. And even some of those are not necessarily following the ins and outs of Brexit and its effect on Ireland. There would be a few who are following it, and they would be generally supportive of my position. But for 97, 98 per cent of my constituents, there isn't really any sort of interest. Perhaps 30 or 50 years ago there was a benefit to it. There is neither a benefit nor a penalty today.'

Contrary to what might be perceived in Ireland, he is very clear that there is no such concept as an Irish voting bloc. Not in his district in Philadelphia, Pennsylvania, and not nationally.

'You do have a relatively small number of people who would be following current Irish issues and might be politically active on those issues. In the grand scheme of the overall population, though, that's a relatively small percentage. Irish Americans today are a large and diverse group ideologically, socioeconomically, even in terms of religion. Irish Americans are much like any other white American group in that they're split pretty evenly between the two parties.'

And he's seen that split in his own district – his home state of Pennsylvania was one of the Democratic-leaning states that flipped and voted Donald Trump into the White House in 2016. The state had voted Democratic in every presidential election since 1992.

'You might have someone who is a fourth-generation American, mostly of Irish descent, but probably also some other ancestry. That individual might be living in an affluent suburb, might be an attorney, and leans Republican. You might have another Irish American who may be second-generation, in the building and construction trade, and really cares about labour issues and leans Democratic. There's a lot of diversity in the true sense of the word within the 40 million Americans who identify as having Irish ancestry.'

And he has a bugbear about the Irish American press and the media in Ireland commenting that Irish America is turning conservative. He doesn't agree.

'I see a lot of commentary related to the number of people in the Trump administration with Irish surnames and a lot of think pieces about Irish America being more conservative. I think that is a very simplistic and narrow view, and here's why. If you take the last two administrations, the Obama administration and the Trump administration, they both had a large number of individuals with Irish surnames in them. Obama, I remember, even made a reference at one of the St Patrick's Days at the White House.[10] So that very much tells you the state of Irish America. And I would expect the next administration, whether Democratic or Republican, to probably follow that trend.'

Boyle was first elected to the US Congress in 2014 in the 2nd district in Pennsylvania, which takes up much of the city and suburbs of Philadelphia. The joke said of Pennsylvania is that it is two big industrial cities: Philadelphia on the east of the state and Pittsburgh on the west and Alabama in between – meaning very conservative and very Republican. It has one of the highest numbers of National Rifle Association membership of any state in the US.

Boyle says his district is one of the most Irish in the state and, given where it's located, it's one of the most Irish outside Massachusetts and New York. 'I probably would have one of the larger concentrations of people of Irish background. But even at that it wouldn't be any more than 12, 15 per cent of my district. And within that 12 or 15 per cent you will get some who vote just based on the name. It's not nearly enough to win any sort of election – God knows, my elections would've been a lot easier if that were the case.'

10 See 'The White House' (page 154) on the shamrock ceremony for shout-outs to Irish Americans in the respective administrations.

Notwithstanding that there is not an organised, cohesive and persuasive voting bloc that could get him elected, Congressman Boyle does point to the Irish American Democrats organisation as a group that did benefit him in some ways. There is an Irish American Republicans organisation too, but they are not as well organised, according to Boyle, as the Democratic group.

'I would say Irish American Democrats was important. It was one group of probably a couple hundred that are heavily involved in Democratic politics and fundraise for candidates. To give you an idea of scale, generally they're able to write one or two thousand-dollar contributions. If you're not familiar with American politics, one, two or even five grand can sound like a lot of money, but it's part of a campaign that could be costing two million dollars in total.' Every little helps in the costly business of running a US political campaign, but the contributions the Irish American organisation can make to congressional elections – or presidential elections, for that matter – are not going to swing too many voters or buy much media broadcast time.

And there is definitely no national Irish American vote, he says, that would be capable of winning a party's nomination for a presidential candidate or a presidential election itself. 'I wish there were, though,' he jokes. 'Perhaps you'd be seeing Boyle 2020,' hinting at something that others whisper about in his future.

The reason there is no unified Irish vote is not just to do with the Irish and their assimilation into American society, but also to do with how America votes in general.

'First off, the "Irish vote" is not a monolith. You have Americans of Irish descent who are politically conservative, Americans of Irish descent who are politically liberal, and all different ideological views in between. And secondly, in the context of overall American politics and society as it stands in the early twenty-first century, our politics tend to be more racially polarised and not as much ethnically polarised.' In other words, there are black voters, white voters, and an ever-increasing number of Hispanic voters (often referred to as 'brown' voters). Polls and statistical analysis tend to break down voters along these lines – African American/Hispanic/ Caucasian – and not according to ethnic origin. Because Irish Americans are mostly white, they are counted as white voters, alongside all sorts of other hyphenated Americans, and their individual views are lost in that collective.

But it was that emotional and personal connection to Ireland, that pride in where he came from and what his father and grandparents had done, that moved Brendan Boyle to seek out the Friends of Ireland caucus on Capitol Hill once he was elected to Congress. He didn't wait for them to come calling.

'The fact that I am the only member of Congress now who has an Irish-born parent, I felt a personal obligation to make sure that I was active in the group. But generally, my involvement on these issues is because of my own background and my own strong personal interest. There are some people in my district who feel similarly and are appreciative of the efforts. There's almost no one who would be opposed to my involvement in these areas. But the vast majority of my constituents would otherwise just be ambivalent.'

But from a political perspective, Boyle says there's no question about the value and power of membership of the Irish American caucus. 'There's no question about it – I mean, so many other similarly sized countries to Ireland look enviously at the sort of access that Ireland has on Capitol Hill and with presidential administrations. And it's something I'm very proud of – it's a great Irish American success story that this group can end up having so many taking a strong personal interest in Ireland, whether it was a generation ago, or five, six generations ago. That attachment gives Ireland great influence here.'

That influence is down to the congresspeople and senators who are part of the Friends of Ireland caucus. Originally it was an adhoc committee of what one long-time congressman described as 'the hardliners'. Separately, there was the Friends of Ireland group with Senator Daniel Moynihan and Senator Ted Kennedy. But after the Good Friday Agreement, with peace looking to be secured, they merged to become the Friends of Ireland caucus. When the Republicans hold the majority in the House of Representatives, a Republican is the chairman – most recently Congressman Pete King – and the Democrats hold the co-chairmanship. When the Democrats hold the majority, as they do now, that party also holds the chairmanship, which in 2019 is Congressman Richie Neal, and Congressman King has become the co-chair. To a man and woman, those who've been members of the group speak of the experience fondly. Wearing their party colours, they can engage in blazing rows on the floor of the House, in the media, in the corridors of Capitol Hill, but once they attend a meeting it's all about the green and a strict bipartisan spirit.

As the current co-chair Congressman King described it, 'The Irish are always fighting with each other. They actually act as peacemakers in the US, so being Irish brings us together.'

As individuals, they also maintain close ties with the Irish embassy in Washington DC, and it is down to the hard work of successive rotations of Irish diplomats that those relationships are forged and bound. The payoff for visiting Irish politicians, of whom there is often a steady stream, is that they can have high-level meetings on Capitol Hill when in town. Often these high-level meetings have very low-level agenda items – items which have little to no political value or even interest for the US politicians. But is it an obligation the Irish American politicians feel or do they enjoy these meetings?

'I like to do it,' says Boyle. 'Then again I am conscious that I am the only member of Congress with an Irish-born parent. I personally feel a certain responsibility to make sure that I take those meetings. But it's not just taking the meetings for a picture and talking for fifteen or twenty minutes. It actually is about working on certain issues.'

Time and again, Boyle returns to that fact that he's the only member of Congress with an Irish-born parent. He had a cross-border compatriot in New York congressman Joe Crowley – his mother was born in County Armagh – until January 2019, but as he's lost his seat, the mantle rests squarely on Brendan Boyle's shoulders. He fights hard for Ireland where he can but is aware that, with declining immigration, the lack of fresh blood is having an impact. 'There probably wasn't only one member of Congress with an Irish-born parent one generation ago, or two or three generations ago, so that shows the way in which immigration has dried up.'

Boyle is concerned about the future. 'It hasn't affected Ireland's influence so far, as the month-long St Patrick's Day events show. But it is an open question what sort of effect it'll have twenty, 30, 40 years from now. I hope and pray that I'm not the last member of Congress with an Irish-born parent – that's something that I really do worry about.'

He continues. 'Now, today, in Congress you have people who might be third-, fourth-, fifth-generation Irish American who really care about that identity, and they're very involved, they're members of the Friends of Ireland caucus. They give me hope that this can be sustained. But it is something that I worry about. Will that influence wane over time if we don't restart that immigration pipeline?'

He thinks it is something that the Irish government is worried about too. 'From the Irish government's activism of the last several years, it's pretty clearly a major priority for them too.'

Boyle says he and other fellow Democrats and also Republican members of Congress are often invited to the Irish ambassador's residence in Washington DC for dinner and to talk about immigration reform. It's this bipartisan cooperation that many current and former members of the Friends of Ireland caucus point to when praising the point of the grouping. It offers a way to reach across the political aisle, or even just have a civil conversation behind closed doors, at a time when relations on the floor of the House or the Senate are fractious.

Brendan Boyle points to his long-time friendship with the acting chief of staff to the president and one-time South Carolina congressman Mick Mulvaney – another member of the Irish American caucus.

'We became friends through the Irish caucus. Mick and I decided that we would honour the heroes of 1916 by getting a tree planted on the Capitol Hill grounds, an honour that generally goes to Americans and very rarely goes to another country.'

Then Taoiseach Enda Kenny also took part in the ceremony in May 2016 that was organised to commemorate the 1916 Rising. There are very strict rules governing the planting of trees on Capitol Hill. There were no commemorative tree plantings in 2015, and the most recent one before that happened in 2014 to honour Anne Frank. Tree plantings to mark friendships between the US and other nations happen even less frequently. Before the 2016 event, the most recent took place in 2007 to mark US–France relations.

Both Brendan Boyle and President Trump's chief of staff Mulvaney speak fondly of this initiative and their close work together.

'We became friends as part of that, and we're still friends today, even though he's now the chief of staff of an administration that I've been extremely critical of. The Irish connection builds bipartisan relationships during an otherwise pretty divided time politically here in the US. Mick is someone I get along with quite well, and I don't agree with on most political issues. But he's someone I can comfortably call a friend and have that relationship. He's a conservative from South Carolina; I'm a Democratic member from Pennsylvania. The odds that we would have come together and been friends without that Irish connection are pretty low.'

The tree that Boyle and Mulvaney planted still stands on the grounds of the Capitol for all to see.

———

In the 1970s, Richard Neal, then in his twenties, was coming of age in Springfield, Massachusetts. As might be expected for someone born in the most Irish state in the US, he has Irish blood in all strands of his DNA. One grandmother was born in Banbridge, County Down. His mother's family were all from Ventry in County Kerry and father's grandmother was born in Roscommon.

Growing up in Springfield, he says, his Irishness was 'very apparent. I grew up culturally Irish.' His family and friends were all Irish and lived in the Irish neighbourhood. Being Irish wasn't about being aware of, or identifying with, the Troubles: it was a cultural, emotional feeling.

'The John Boyle O'Reilly Club and many of the local taverns had a very noticeable kind of Irish influence. Many of them are located right on Chestnut Street and across Hungry Hill, places like that. The boys' club was African American and Irish, and even a bit of a Jewish community there. Those were the pronounced groupings.'

While many Irish American politicians will point to the Irish sense of public service as their 'call to action', like being a firefighter or police officer, US Congressman Richard Neal (Richie to everyone) knows exactly when he realised his vocation.

'The pivotal moment, obviously, was seeing Kennedy the day before the election in 1960.' 'Kennedy' being, of course, John Fitzgerald Kennedy, or Jack Kennedy as those in the environs of Boston still call him.

Richie Neal was eleven years of age and he remembers distinctly staying home from school to see the Democratic candidate for the presidential election come to Springfield on one of his last campaign stops before polling day.

'My mother was long gone for Kennedy and my father was a bit more discerning, but my wider family, they were all for Kennedy! So I stayed home to see him the day before the election and then about 72 hours later he was the president-elect. It was also that flash of idealism that one embraces at that age.'

He felt a sense, as he says the wider community did, that anything was possible for immigrant families. 'The political culture of Springfield and Massachusetts was very Irish at that time.' The power was real and for an eleven-year-old boy the sense of opportunity and possibility was intoxicating.

He has raised his own children to be what he describes as 'culturally Irish'. His daughter and two of his sons live an Irish cultural life, and a third son is less into the Irish scene. But the family listens to the Cranberries and the congressman likes Mary Black. He laments not being able to have a long conversation with his daughter on the phone at the weekend because she was dashing to bring his granddaughter, Roisin, to her Irish-dancing class.

When it came to seeking political office himself, for Springfield City Council in 1978 and later for US Congress in 1988, unlike other congressional colleagues, not running as 'the Irish guy' was not an option. Not only was he clearly the Irish candidate on the ballot paper, but it was 'a huge advantage', says Neal.

'It's fascinating because the Irish are the largest group in Massachusetts. Even to this day. And they always have the best voter turnout. We [the Irish] were always going to vote.'

He says there was a level of engagement that was not quite the same for other ethnic groupings. 'I think it came from the influence of the west of Ireland, which was where my constituents all come from. It was based in those conversations that took place in rural

Ireland where they were very politically active, and they brought with them this interest in the civil service. The first generations were labourers and it's never fully acknowledged, I think, in the way that it should be, that they did many jobs that others wouldn't do. The rates of poverty were really high.'

He says the odds were high of immigrants working up quickly from those jobs to the middle and professional class. 'You start to see the breakthrough between politics and the civil service. You start to see the Irish becoming the cops, the firefighters, the tradesmen, the teachers, and then the executives and into the private sector. By that time, you had witnessed the election of a president, which is always a yardstick for success that a people finally have.'

John F. Kennedy not only had strong Irish backing upon his election, but he was also the first (and so far only) Catholic elected as president of the United States, and Richie Neal believes that can't be understated either.

Neal says that Kennedy, like a lot of politicians, including himself, was charged with a dual loyalty – that he was more loyal to the pope than to the country. 'There was a certain – rather there was a pronounced – hostility to Kennedy's candidacy in large parts of the country because of his Catholicism.'

So for Neal, the fact that Kennedy went to Houston, Texas, and met the Greater Houston Ministerial Council and confronted the issue of Catholicism was 'really courageous'. As a politician himself now, Neal guesses that Kennedy would not have wanted to make that speech, but he did it anyway, because he had to.

'Kennedy's speech that day is magnificent. It's worth reading. He asks them, "did I forfeit my right to be president the day that I was baptised?" It's a stirring speech that he makes, and he gets a standing ovation. He wasn't quite sure how many hearts he had won, but he thought that the presentation was pretty strong. So having seen him in the flesh that day, Kennedy was a big thing in our house.'

That youthful idealism in Richie Neal in the early 1960s made one thing clear. 'Politics was the way up. It was opportunity.'

And he had a closer example than his beloved President Kennedy: Congressman Edward P. Boland from Springfield, Massachusetts, Neal's long-term predecessor in the US Congress. 'His parents were from Annascaul in County Kerry. Then here I come, of the Garveys, my family from Ventry, the next town over. It's pretty extraordinary. And there's more than 65 years between us.'

In 1934, at the age of 23, Eddie Boland was elected to the Massachusetts House of Representatives. He was the youngest of four sons born to Irish immigrants Michael and Johanna Boland from County Kerry. They lived in the working-class, heavily Irish American Hungry Hill section of Springfield, Massachusetts. There are conflicting accounts of why it was called Hungry Hill. The most common is that it was full of impoverished, starving Irish immigrants. The other is that its sloping topography reminded the Irish immigrants of a place they had come from – Hungry Hill in Bantry Bay, West Cork.[11]

Boland served in the US Army for eighteen months in the South Pacific during World War II and was honourably discharged with the rank of captain. He was first elected to the US Congress in 1952 and then re-elected for seventeen consecutive terms. In all he spent 54 years in public service, all the while living in Springfield with his wife, Mary Egan, and their four children, Martha, Edward, Kathleen and Michael. He fought for equal and human rights for the marginalised, the poor, the elderly and those with disabilities. He went to Selma during the civil rights movement to march with Martin Luther King. He was author of the Boland Amendment, which was at the heart of the Iran–Contra scandal in which President Reagan authorised US arms sales to Iran and his

11 O'Connell, James (ed.) (1985). 'The Hungry Hill Neighbourhood'. Published by the Mayor's Community Development Department and the Springfield City Library.

aides at the National Security Council diverted sales profits to the Contras, acting in defiance of the amendment. His legacy is writ large across the US Congress and American public life.

Being 'the Irish guy' helped get Boland elected and, says Richie Neal, the same benefit is still there for him today, even if the demographic makeup of Massachusetts is changing.

'I derive a benefit because, I think, a lot of the Irish immigrant community has become more conservative as time went on. There was no such thing as a Republican who was Irish when I was growing up. It was very much union households, Democrats, FDR supporters and that's not true any more. Now I think a lot of them still keep D next to their name, even if they vote Republican and they vote for me [a Democrat].'

He says there are a number of reasons for this swing back and forward from D to R in this Irish city. 'I think there's that loyalty, even in the suburbs around Springfield, even though they've grown more conservative.' He says that conservatism is based mostly on cultural or social issues, rather than economics. These voters are still proud trade union members, and believe in welfare and looking after the poor, but at the same time they are religious and a lot of the cultural conservatism, he feels, comes from the homily they hear on a Sunday, particularly on issues like abortion and gay marriage. Issues that still divide regardless of the law of the land.

But he also thinks that, as immigrant communities move further from the generation that immigrated, the views and outlook change. 'By the third generation, a lot of other Americans don't look back at their old country other than with a certain sentimentality. But for Irish Americans, it's different. It's because of the difficulties on the island and the experience that they have because many feel their family has been driven off the island. It was a primordial grievance.'

This primeval attachment to the land they hailed from is no more evident, says the congressman somewhat morbidly, than in the obituaries pages. 'It's always amazing when I read the obituaries back home – daily, almost – for those who were born in Ireland because there's still large numbers of them and they carry influence. When I read them, they all say "were children of Irish immigrants" or "were proud of their Irish heritage".'

So being seen as 'the Irish guy' is one thing, but does he need to actively campaign on Irish issues to actually get the votes? And what are Irish issues, in his view?

Voter issues are voter issues, says Neal, and he is an old school politician who takes hand-shaking, smiling for photos, old-fashioned 'pressing the flesh' seriously. But when it comes to what is euphemistically called the Irish issue, or as it was in the UK (and possibly still is now in the era of Brexit) the Irish Question, getting involved in an effort to bring peace to the island of Ireland did win him votes in the eighties and nineties, and still does now. Although he stresses that is not why he got involved.

'It wasn't popular when I got involved. I took the positions I did back when many others wouldn't go near it. There was some criticism in Irish circles; people saying that the issue couldn't get solved. We insisted it was a human rights issue. There was a convergence of issues that happened simultaneously, but part of it was also international events. You had the collapse of the Soviet Union, the Berlin Wall came down, apartheid ended, and here's Belfast. And people would say – well, what about Ireland?'

Neal says the credit he got for taking a role is still active currency. Even if voters in his district are more conservative now, they will still point to the fact that he played a part and he cares about Ireland. And after all, even from the gleaming ivory dome on Capitol Hill, all politics are local.

'The first coffee hour I ever did was at the John Boyle O'Reilly Club. The women that put it on – one was Ann Baker, from Kerry, another was Kathleen Murphy, from Mayo, another was Ann McCarthy, and another was Connie Powers. Connie's parents were both born in Ireland. It really wasn't a coffee hour – it was more alcohol, you know – and after I laid out all these things I was going to do to improve the world, I looked around and I said, "Please, any questions?" A woman says to me [and here he puts on a soft Irish lilt], "How's your grandmother?"'

So Irish Americans are still a very large group and Neal says he does still derive a certain benefit from their support. While he would describe the local voters as something of a cohesive voting bloc for him, he's not sure it is cohesive for Democrats nationally any more. 'I think it is still there for some of us as candidates on an individual basis, but is the vote monolithic any more? I don't buy that. No.'

Like others, he thinks it faded quite some time ago. 'The Irish all voted for Kennedy, then they voted for Reagan. And it wasn't just because of his name – it was more that they had begun the shift, the movement of the Irish away from the Democratic Party. In particular, the movement of Irish Catholics away from the Democratic Party.'

—

Along the coast between Brendan Boyle and Richard Neal, but across the aisle politically, Pete King, the fourteen-term congressman from New York, Long Island, the 2nd congressional district, has a similar tale.

He has Irish grandparents on both sides. His father's mother was born on Inishbofin. His father's father was born in the US to Irish parents, who brought him back to Ireland at the age of four or five, so he was raised in Ireland. His mother's mother was born in Limerick. He grew up listening to stories from Ireland, but mostly all from his mother's side. He says his father's parents never really spoke about it and 'didn't have very romantic views of Ireland'. He adds, 'Inishbofin is kind of a stark place,' and chuckles. They weren't anti-Irish, he says, but they didn't talk about it very much.

But his maternal grandmother talked about Ireland all the time. 'She had more brothers and sisters over there than she had here in the United States, and they were always writing back and forth. She left somewhere around 1910, I guess. She went back, ironically enough, I think, in the summer of '22, during the Civil War. Then she didn't go back again until the 1960s. You know, mainly just for financial reasons. Then she went back a few times in the sixties and seventies. Her brother was in the IRA and he was in jail.[12] I think it was during the Civil War. My mother tells the story of how when they were home in Ireland the family would go down and visit him in jail.'

Congressman King knows this sounds like a movie script – grand-nephew of man jailed during the Irish Civil War becomes US congressman … 'How much of this stuff is true, you don't know – but it is what I was told!' He's also conscious of the risks a senior politician takes in making claims about heritage that he can't stand over, so he elaborates. 'I've actually seen him. There was a portrait of him in an IRA uniform, which I think was done somewhere probably in the twenties or thirties. And he had a pension that they gave all the IRA guys.'

He recounts one story his mother passed on to him. She was just five when her mother brought her two children back to Ireland to visit, during the Civil War. 'They were there for the whole summer and my mother talks about putting pillows in the windows at night to keep the bullets out. Now, again, it wasn't something that was bragged about. It was just spoken about pretty casually.'

With so many newly landed Irish in New York who had fled the country post-Rising or during the Civil War – or afterwards

12 Her brother was in the anti-treaty Irish Republican Army in the Irish Civil War, not the modern-day IRA, classified as an illegal organisation in Ireland and a terrorist organisation in the UK.

because they had picked the losing side – it's inevitable that the divisions and battle lines would be drawn in their new home. Pete King's grandmother was one such, as was his mother, although she was only a child at the time.

'My mother's mother was very much an Irish nationalist, very anti-British. She loved de Valera, and had no time for Michael Collins. She was very proud of the fact that when they lived in Manhattan, they lived on the East Side and they were in St Agnes parish where de Valera was baptised. I used to hear that quite a bit.'

He was in no doubt as to her views. 'My grandmother also said, which is probably not politically correct these days, that they should have shot Collins before he went to London, not when he came back.'

Nothing of the sort was discussed by his father's side of the family. Just to keep things interesting – or, as Pete King says, 'to add to the confusion' – his grandfather had been an Episcopalian and converted. 'But it was never really a true conversion. It was a conversion of convenience because they were always arguing about England and Ireland and Catholics and Protestants.' So it was a colourful and lively Irish upbringing, and he feels because of that he was probably more aware of Irish politics than 'the average Irish American kid at the time'. And it is this upbringing that he pinpoints as the reason that he took such an active role in Irish affairs as a politician.

He was already 'in politics', as he puts it, when the civil rights movement started in Ireland. Although he wasn't elected to any office until 1977, he was already a political activist, campaigning for others and organising in Long Island. He says it struck him at that point that a lot of other immigrants and ethnicities in the US were helping out with conflicts around the world. 'Obviously, a lot of African Americans got involved in South Africa. Later, there were a lot of Polish Americans getting involved with Solidarity. Certainly, you had Jewish Americans involved with Israel.' It seemed natural to him that Irish Americans would get involved in what was going on in Ireland, North and South.

When he first began running for elected office, he thought about the impact being 'the Irish guy' would have – would it translate into electoral support? And as he seeks re-election to the US Congress every two years, does it still?

'It probably did at one time. But that, I think, was always exaggerated. I mean, people generally would have supported what I was doing, even though the local media would attack me. But it was not a raging issue like it would have been in the 1920s or something like that.'

Congressman King says he has 'never campaigned as the Irish candidate.'

He feels the concept of an Irish voting bloc is a diverse one. 'It's not an Irish voting bloc which has very much to do with Ireland any more. It would be almost a demographic thing. Generally, older Irish Americans, I'd say people over 40, we're more socially conservative. We're also more patriotic or pro-American or more conservative on foreign policy. The culture of Irish America on many social issues would be more right of centre. For example, it was the Irish voting bloc, if you want to call it that, who voted overwhelmingly for Ronald Reagan. They may have voted for John Kennedy, but they also voted for Ronald Reagan.'

He also believes the Catholic–Protestant divide that existed in Ireland when many of the immigrants left in the 1920s was alive and well when they arrived in New York.

'That was the general sense, though I always emphasised this was not a religious issue – it was just a uniform that people wore. I mean, you were branded as a Catholic or branded as a Protestant, which was a line of demarcation, but the days of theological fights in Northern Ireland were long gone. This had nothing to do with that. It was just a way of identifying. A person who was Protestant was basically pro-British. A person who was Catholic was basically pro-unification. That was it.'

King explains his rationale for the fervent support – financial and political – for the Irish Republican movement in New York. 'The core of the Irish Americans who were here, the Irish immigrants who came in the twenties, many of them were republicans. There are very few pro-British Irish organisations. I don't know of any pro-unionist groups in New York. Among the Irish organisations over here, most of them were founded or started or organised by people who came over here in the 1920s and they were very pro-republican, if not IRA.'

In acknowledging that he only knows Irish Americans of a nationalist or (Irish) republican persuasion, King has hit on something that has posed problems for diplomats in Northern Ireland – how to explain to a local mixed constituency that the population in the US is not mixed.

Norman Houston has been director of the Northern Ireland Bureau in Washington DC for almost two decades. As Northern Ireland is officially represented at a diplomatic level by the British embassy and the British ambassador, he cannot be described as an ambassador, but he does similar work for Northern Irish interests.

He says that, in his experience, when people talk about Irish America, they are talking almost exclusively about Irish Catholic nationalists. Those of a unionist tradition, with an Ulster Scots background, do not fall easily under the Irish American umbrella.

'The truth of the matter is that they don't exist in what I call a political position. Those from the older tradition, who've been here longer – the Scots Irish – came almost exclusively before the Revolutionary War. That's not to say they don't have an interest in Ireland or Northern Ireland, but they're not a political force.'

He says it can be problematic explaining this to some of those at home in Northern Ireland. Americans who have ancestors that left the six counties several generations ago have chosen to identify as Irish, or Irish American. In his experience, they don't identify as *Northern* Irish American. For them, he says, Irish American is a word they use for the island of Ireland, rather than a pejorative view of any other identity in Northern Ireland.

But while Irish America may be predominantly Catholic and nationalist, Houston says that does not mean there isn't a place for the unionist tradition. 'In my view, there's always been an anxiousness to make sure that every shade of political opinion in Northern Ireland was welcomed to Capitol Hill. Maybe they dream of a united Ireland at some stage, but they don't roll that out. There's always a willingness to listen to a unionist perspective.' He says that looking for peace and reconciliation has trumped political divisions.

—

Congressman King says that when he looked at Ireland in the late 1970s and on into the eighties, he had a deep feeling that the US was uniquely placed to try to broker some sort of an agreement.

'The more I looked at it, it seemed like all sides were afraid to make the first move, that they'd be left out. The Irish republicans would still be talking about 1922 and the Treaty of Limerick while the British thought they were dealing with terrorists. Why should they sit down with them if they would just take advantage? So I thought the US could be an honest broker.'

But not only did he feel there was a part for the US to play – rather controversially, he also felt that the IRA needed to have a legitimate role to play in any negotiations. Only, he says, 'because they had such strong support on the ground and there were very serious human rights violations.'

This was the opposite view to that taken by the so-called Four Horsemen – Senator Ted Kennedy, Speaker of the US House of Representatives Tip O'Neill, Senator Daniel Patrick Moynihan and New York Governor Hugh Carey. According to how Congressman Pete King sees it, 'they basically spent all of their time denouncing the IRA and making John Hume out to be the only spokesman for the North.' He says this was the same during the Reagan administration, and the bipartisan cooperation between Tip O'Neill and President Reagan was also about ignoring the IRA. 'So I felt that it was important to say that the republican movement had a real role to play.'

During this time, Congressman King paid many visits to Ireland. He was in Northern Ireland in December 1980, during which time the number of prisoners on hunger strike was increasing, as each week more prisoners joined the original seven who had started in October of that year. He was back in 1981 for the Plastic Bullet tribunal and made repeat visits in 1983 and 1984. He was asked by the republicans (Irish, not US) to be an observer for the so-called supergrass trials.

He points out that he had contacts on both sides of the community at that stage. While deeply connected with the republicans, and the IRA, in 1984, he says, loyalist families asked him to go over, and he was then the subject of a documentary made by Channel Four in Britain.

He said the key for him was that when he went to either of what he describes as the 'extreme' sides of the community, he met people who were 'very normal, not crazies'.

'These were not fanatical people, and the fact that even then they were getting more votes in elections, to me, showed they were not a fringe. Again, when you were there with them, they were very much like Irish American families I had grown up with, and that's why I became an advocate for them.'

But while he was known as something of an IRA sympathiser during this period,[13] Pete King says his judgement was never clouded. 'Once we did have the IRA ceasefire, it demanded an

---

13 See, for example: Shane, Scott. (2011). 'For Lawmaker Examining Terror, a Pro-IRA Past'. *New York Times*, 8 March 2011. <https://www.nytimes.com/2011/03/09/us/politics/09king. html.> Moloney, Ed. (2010). 'Rep. King and the IRA: The End of an Extraordinary Affair?' *New York Sun*, 22 June, 2005. <https://www.nysun.com/national/rep-king-and-the-ira-the-end-of-an-extraordinary/15853/>

end to all violence, so when the IRA broke the ceasefire in '96, I was one of the first to denounce it.' He knows that his reputation preceded him too. When he first came to Congress in 1992, he says, the Democratic congressman and one-time Speaker of the US House of Representatives (and another proud Irish American) the late Tom Foley was telling people to watch out for him because he was 'a friend of terrorists'.

King also worked hard – mostly behind the scenes – pushing towards the Good Friday Agreement. While the work of Senator George Mitchell, the official US envoy, is well documented, King continued his work as a behind-the-scenes advocate, particularly with the IRA. He has written a just-about-fictionalised account of the process in a novel (his second) called *Deliver Us from Evil*. He says everything in the book is accurate, but he has combined some of the characters to make them composites because 'I didn't want to put anybody on the spot.' The book's central character is a Republican Congressman from Long Island, NY called Sean Cross, a hero who works to stop a conspiracy to derail the peace process, when an IRA informer in his district is murdered. The fictionalised Congressman Cross enlists his old ally President Bill Clinton to help solve the murder. According to the publishers: 'the story moves back and forth from New York, to Belfast, to London, to Washington, DC, while real life figures such as Gerry Adams, Tony Blair and Bill Clinton move seamlessly in flashbacks from actual historic roles to fictional exploits.'

The book is set in 2006 and features this discussion between the lead character Congressman Cross and former President Bill Clinton in a New York restaurant as they discuss the US role in the peace process.

'Sean, looking back on it, do you think I would have been able to move on Ireland the way I did if the World Trade Center had been attacked in 1991 instead of 2001?'

'No. And I've given that a lot of thought.'

'The terrorism aspect.'

'Yeah. There was such an outcry against terrorism after the Twin Towers, it would have been almost impossible to distinguish the IRA from al-Qaeda — even though to me there was no comparison.'

'It would have been very tough.'

King describes himself as a 'go-between' for the White House and Gerry Adams, particularly at a time when it was politically unpalatable for anyone in the administration to be seen to be dealing with Gerry Adams, leading ultimately to the decision by then president Bill Clinton to grant Adams a visa to the US – and the subsequent decision to take it away and ultimately restore it after the 1996 bombings.

King was so grateful for the role played by President Clinton that he was one of only four Republicans to vote against all articles of impeachment charging Clinton with obstruction of justice, perjury and abuse of power.

In April 1998, when the Good Friday Agreement was signed, King again resumed his role as go-between. The deal was meant to have been signed on the Thursday and be known as the Holy Thursday Agreement, and King jokes that he was telling everyone 'you know the Irish never get anywhere on time – it won't be done on Thursday'.

He says he spent a lot of time in Ireland again during this period, dealing with both loyalists and republicans. But 'not too much with the Unionists. They can be difficult to deal with.'

That role of 'honest broker' played by the United States in the long decades of the peace process leading up to the Good Friday Agreement is often highlighted as the peak of any sort of meaningful Irish American relationship. There was something real to play for, the stakes were high and there was arguably a clear need for careful negotiation and diplomacy from outsiders free from emotion and historical perspectives.

But King is clear on what ultimately made a difference. 'The true tipping point was Bill Clinton getting elected.'

He explains. 'That was basically it. Bill Clinton treated this as a legitimate issue, not just as a fringe issue, not just as psychopaths killing innocent men, women and children. He would understand the social reasons for it, and also the fact that it was a war that could go on forever. No one would win, no one would lose. I think that was pretty much Gerry Adams's thinking too. I mean, he realised the IRA, if they wanted to, could fight for ever without being defeated, but what purpose would it really serve? On the other hand, if you just gave up, that wouldn't serve a purpose either. So you had, I think, the perfect alignment of the stars. You had Bill Clinton. You had Tony Blair. You had Bertie Ahern. And you had Gerry Adams and McGuinness. Trimble was sort of a weak link there in that he never really got his people to come around. I mean, Adams was pretty tough. Adams did what Arafat never did. He really told his people that this was it, that the agreement was going to be the agreement, and put himself at risk with hardliners in the IRA.'

As do almost all those involved at the time, King pays great tribute to John Hume. 'I've got to give John Hume tremendous credit because once Adams started getting a visa, he'd come to the US. I'm using this trivial expression, but Gerry became the flavour of the day. At every St Patrick's Day event, there would be 40 reporters and photographers standing around Gerry Adams and John Hume would be by himself. Adams was aware of that, and he was always trying to make sure Hume was part of it.'

King plays tribute too to former President Bill Clinton. 'I supported Bush in '92, but I think if George Bush had been in the White House, if John Bruton was the taoiseach and Margaret Thatcher was the prime minister, I don't know if you would have had a deal. I had leaders from the North and from the Irish government telling me that Bill Clinton understands the psyche of the North better than many of the Northern leaders. The one person he couldn't quite figure out was David Trimble, but he understood the loyalists, he understood Hume, he understood Adams. He understood the reaction of people on the ground. He understood, also, why the Unionists would dig in. They were afraid of what was going to happen. They were afraid of losing everything. Clinton was aware of all that. It was really interesting how, even when talks would break down at Stormont, everybody would come to Washington on St Patrick's Day and they would all meet with Clinton individually and then they would have these side meetings over here. So they were able to do things in Washington they couldn't do in Stormont. I mean, some of these guys were like travelling salesmen – they were showing up in Washington more than they were in Northern Ireland. Guys like David Irvine, you know, some of the fringe players, you could say. After a while, even Paisley Junior would come. So I really give Clinton credit for that, for bringing people in from the cold.'

As the congressmen have all pointed out, the years between now and then have – from a US perspective, anyway – seen relative calm. No bombs, no hostages, no victims on the nightly news feeds. This has also occurred at a time when US news coverage – broadcast, print and latterly online – has become more domestically focused. More coverage of US domestic politics and, sadly, higher incidents of domestic terrorism and mass shootings mean the bar for 'foreign news', outside conflict zones involving US troops or the imminent threat of nuclear war, is a lot higher. Irish and British politics don't cut the mustard. At least, they didn't cut the mustard. Then along came Brexit.

Brexit has had a life-extinguishing impact on politics in Northern Ireland. At the time of writing (July 2019) the institutions have been frozen for more than two and a half years and are showing little chance of resurrection as long as the DUP remains a pro-Brexit party, keeping the Conservative British government in power. The British government, therefore, cannot be a neutral negotiator and Sinn Féin cannot expect transparent negotiations.

Pete King thinks resurrecting the institutions on this occasion is not a place for US involvement, despite (as yet unfulfilled) promises from President Trump to appoint a US special envoy to follow in the footsteps of George Mitchell, Gary Hart and others.

'There's only so much outsiders can do on this. I don't know how much more we can do as far as getting the government in the North back up and running. It really is time for them to start doing it themselves. If there is particular help they want us to give, I'm sure the US will give it.'

—

There is a place called Tipperary Hill in Syracuse, upstate New York. It's so called for obvious reasons – most of the original inhabitants came from Tipperary. Everyone in the neighbourhood at one point was either Irish or had a connection to Ireland. Names like Collins and Ryan percolated the telephone directories.

At the corner of Burnet Park Drive and Tompkins Street, there's a traffic light on the hill. But it is not styled like a traditional light sequence with red on top, amber or orange in the middle and green on the bottom. This Tipperary Hill traffic light has the green on top, then the orange and then the red. The green is always on top.

There is a story told locally about this traffic light. It goes that when the light was first put up with the green on top, in deference to the Irish community, a state transportation engineer said it had to be changed because it was against standards. So the 'upside-down light' was taken down and another put up with the red at the top in the usual way. A local group on Tipperary Hill who called themselves the Stone Throwers would then go up at night and throw bricks and stones at the light signal until it was smashed. The city officials would then come and fix the light. This happened three or four times until the city realised, as the story goes, that they couldn't beat the Stone Throwers. They gave up and put up a set with the green light on top again.

One of those Stone Throwers was a man called George D'Arcy, uncle to much-decorated US Congressman Jim Walsh, who grew up in Tipperary Hill. Jim Walsh laughs every time he tells this story, and he has taken many Irish and Northern Irish politicians to these traffic lights to have their photo taken in front of them.

As a former Republican US congressman, Jim Walsh is not a diplomat in an official capacity, and he's not a member of the foreign service. But, like many in the US, he works for Irish affairs, and is considered an unofficial ambassador. He was one of those who first got President Bill Clinton involved in Irish affairs and although he has now retired, he is still involved himself. He's one of the co-chairs – along with Democratic former congressman Bruce Morrison of the Ad Hoc Committee to Protect the Good Friday Agreement. He says that, growing up, he had an awareness of Irish America but he did not have an awareness of Ireland.

On his father's side, Walsh's great-grandparents David Walsh and Mary Hughes were married in 1870 in Killala, County Mayo. They had eight children born in Ireland, including Walsh's grandfather Michael J. Walsh. The great-grandfather moved to the US in March 1895 with the eldest son, and his wife Mary and the other children, including Walsh's grandfather, came a month later. They all arrived through Ellis Island and settled in upstate New York.

Jim Walsh's grandfather was nicknamed Silent Mike, and apart from a few stories about Killala, he didn't talk much about Ireland – or anything else. 'It was a kind of a distant, misty, unhappy place for them. They were assimilating and trying to become Americans as fast as possible.'

The former congressman has less information about his mother's side other than they were D'Arcys from Nenagh in County Tipperary, and it was his great-great-grandmother who travelled over. Jim Walsh's wife's maiden name is Ryan and she traces her history back to Tipperary too.

Unlike some neighbours in Tipperary Hill, including Terry McAuliffe, who would go on to be governor of Virginia and a close adviser to President Bill Clinton, the Walshes were not Democrats but Republicans. Walsh's father was mayor of Syracuse and a member of Congress, he himself was a member of Congress, and his son Ben is now mayor of Syracuse. However, in an example of how the area has changed, Ben Walsh ran and was elected as an independent.

Walsh doesn't think the Irish ancestry had anything to do with his family's call to public service – a brother and sister of his are judges – but he thinks being Irish helped their Republican family with their Democratic neighbours.

'I think it helped being an Irish American Republican because Democrats would cross the aisle for you and vote for you. Irish Americans tend to be more Democratic than Republican. We would always do pretty well with the Irish American sections of town, even though they were Democrats.'

Walsh says New York city and state has changed a lot in the last 40 years.

'Up until probably the 1980s, the city was still Republican. But today it's not, it's overwhelmingly Democratic and our home county, which was a solid Republican county, is now majority Democrat, marginally.' The politics have changed pretty strongly from Republican to Democrat, but he says it is more moderate politically in upstate New York than elsewhere.

The swing from moderate Republican to independent and Democrat is something Jim Walsh thinks is happening at a national level, and has nothing to do with Irish Americans – many of whom are actually swinging in the opposite direction.

'I think the Republican Party has become more and more a Southern, conservative-based party on the national level. And people don't identify with that where I live. They're a little more moderate in their political views. I mean, I wouldn't call them liberal but, you know, moderate, centre-right, centre-left. But those folks used to be Republicans and now they don't really identify with the party. New York is the home of Rockefeller Republicans and there aren't a lot of Rockefeller Republicans left.

'The Teddy Roosevelt and Nelson Rockefeller tradition was that government can and should do some things to make society fairer and at the same time be pro-growth and pro-business. A conservative view of government is: our friend can be our friend but don't get in the way of business. That really has changed and the people who are in the Republican Party now tend to be the conservative base of the party, and of the community really. And the Democrats haven't really grown nationally. In the cities where the minority population is larger, they tend to be Democrats. Most people have gone outside of the two-party system to enrol as independent.'

He thinks that one-third of the electorate in his home county in Syracuse, once a Republican stronghold, are now aligned with neither party and are independents. That is a trend he has noticed over 25 years.

Jim Walsh ran for Congress ten times but says it was only when he was running for his third term in the US House of Representatives that Irish America became part of his campaign. This, he explains, was because he suddenly had a national profile that had come from his involvement in Congress with the peace process, granting a visa to Gerry Adams, the trip by President Clinton and a congressional delegation in 1995. He says that all started to 'play well for him' as a leader in the Irish American political scene and locally for him too. And anything that can be converted into political capital is good for those who have to seek re-election every two years as the members of the US Congress do.

'I think mostly it was a positive. There were some people who did not like my involvement in foreign affairs or specifically in Irish affairs, and what they considered to be internal UK politics. I did get some criticism for interfering in Northern Ireland because it was part of the UK. But there were far more people who said "Yeah, that's great. There's injustice there. Be involved. Help out."'

From the pre-Good Friday Agreement days when he was one of the cross-party congressional leaders who played a role in the negotiations, through his retirement in 2009 and until the present day, Jim Walsh says he's 'never stopped' trying to work for Ireland. He's been to Ireland 25 times in that period, making friends in Belfast, Dublin and London.

'I got to know people in all the parties. I've kept the dialogue open with them. There are very few Republicans in the US involved in this. I am one of the very, very few. But if you want to be successful in Washington, it's much easier to do it on a bipartisan basis. I've always been fortunate, because with my Republican label, I'm always included. It resonates far better when people can say "Bruce Morrison and Jim Walsh leading this Ad Hoc group, one a Democrat, one a Republican", because both parties are represented.'

At this point, for Jim Walsh, Irish issues are still a 'labour of love' for him, but he does lament that there aren't more Republicans involved. While there are Irish American Republicans and there is on-the-ground support, that has not manifested as elected officials playing a role.

He thinks it has to do with where the bulk of the Irish population, and the traditional Irish American strongholds, are based.

'The Irish, to a large extent, are urban and Eastern and so those cities are all Democrats. Boston, New York, Pittsburgh, Chicago, they're all dominated by the Democratic Party. Politics reflects local priorities and local values, especially in this country. So if you have a large active Irish American population, you need to represent them and they demand that you be involved.'

He says the Republican Party hierarchy has never made any attempt to reach out to Irish American voters and when he got involved it was nothing to do with local politics – it was to do with curiosity rather than a constituency.

Although he acknowledges that lots of Irish Americans from both parties voted for Donald Trump in the 2016 election, he says that had nothing to do with Ireland. They voted for him based on economic reasons or trade policies, or any number of things.

While the Republican Party may be slow to reach out to Irish Americans, and the Democratic hierarchy may be sliding off as well, this does not mean that Ireland, it's diplomats, politicians and unofficial ambassadors are easing up on their effort and energy levels. Particularly given that many of those who have been most active in pushing Irish influence have retired from elected office, like Jim Walsh, have been rejected from the electorate like Joe Crowley, or are in their 70s, like Peter King and Richard Neal.

Cultivating fresh contacts on Capitol Hill and beyond requires 'a serious amount of legwork', says former ambassador Anderson.

'You can't be complacent, you can't take anything for granted because America is changing and there are competing, of course, sources looking for the same level of access. The whole country is radically changing, the demographic is changing. Irish America is shrinking in proportional terms, so the rationale that lead to us having that level of influence is changing. We are still disproportionately strong on Capitol Hill because of the type of immigration we had and the way our immigrants went into politics and went into channels that lead them to being influential in their communities, but none of it can be taken for granted.'

—

Since the Democratic Party took back control of the US House of Representatives in the midterm elections in 2018, both Congressman Brendan Boyle and Congressman Richie Neal have dragged the issue of Ireland back into the US and, by extension, the international news agenda. They have campaigned hard to force the UK to consider Ireland in its future plans and to make clear in no uncertain terms on whose side the US House of Representatives would stand should it be forced to do so.

It has also had a ripple effect in the UK and in the UK embassy in Washington DC, where there are now regular meetings with the Irish American caucus to keep British lobbying efforts apace with Irish ones.

On 29 January 2019, Congressman Brendan Boyle introduced a bill on the floor of the US House of Representatives. House Resolution number 88 expressed the opposition of the House to 'a hard border between Northern Ireland and the Republic of Ireland'. It pointed out that the Good Friday Agreement had been struck between the governments of Ireland and the UK and the Northern Ireland political leaders and agreed upon by the people of Northern Ireland and Ireland on 22 May 1998. It stressed that the US was one of the guarantors of that agreement and as such continued to support Northern Ireland in many ways, including through the International Fund for Ireland and the appointment of successive peace envoys. The resolution also expressed concerns about the future of the Good Friday Agreement and sustained peace.

The resolution calls for it to be 'Resolved, That the House of Representatives opposes the imposition of a hard border, whether one that is strongly controlled by officials, police, or soldiers, or a physical barrier, between Northern Ireland and the Republic of Ireland.'

Alongside that development, Congressman Richie Neal said that any post-Brexit US–UK trade deal would not make it through the US Congress if it contained any elements that would threaten the

Good Friday Agreement or suggest the return of a hard border on the island of Ireland or in any other way harmed the continued peace. As the chairperson of the powerful Ways and Means Committee, which has the role of rubber stamping any trade deal before it gets to the floor of the House of Representatives for voting, Richie Neal has more influence than many other Irish Americans on what may happen with a US–UK trade deal post-Brexit. 'There is a role for the US here, and we'll play that out if they [the British] want a bilateral trade agreement, and it has to go from this very committee. I've made my position known to the United States trade representative that the border is going to count with us. It has to. You can't do trade agreements with what at that time would be a United Kingdom with the North of Ireland. How would you do that? The border becomes a stubborn issue again.' The US trade representative whom Neal speaks of is Robert Lighthizer. His full name is Robert Emmet Lighthizer, yet another Irish American in a high-profile role in the administration. His father and his son are also Robert Emmet Lighthizer, named in honour of the Irish patriot.

Also, on 5 February 2019, some 40 Irish Americans from the Republican and Democratic parties formed a new group – the Ad Hoc Committee to Protect the Good Friday Agreement. They co-signed a letter to then British Prime Minister Theresa May and Taoiseach Leo Varadkar expressing their alarm that Brexit might be placing the Good Friday Agreement in jeopardy, warning against any harm that Brexit might cause to ongoing peace efforts in Northern Ireland.

The group of high-profile individuals said they were writing to the two political leaders as the 'co-guarantors' of the twenty-year-old peace deal.

One of the co-chairs of the committee, former Republican Congressman Jim Walsh, said the decision by Prime Minister May to reopen the withdrawal agreement with a view to finding an alternative to the backstop 'alarms us'. He added that the political leaders who supported Brexit seem to have 'little knowledge of Northern Ireland'.

In the letter, the 40 senior Irish Americans pointed out that 'Irish America helped to lay the foundations for peace and justice in Northern Ireland,' and that they had a 'growing concern that the GFA may be in jeopardy given the decision by the government of the United Kingdom to reopen the withdrawal agreement.'

Although the Good Friday Agreement is an international legally binding treaty, the politicians and businesspeople expressed their unease at what they view as 'pro-Brexit advocates in London' who 'have set about diminishing the importance of the GFA almost to the point of dismissing it as irrelevant'.

The second co-chair of the committee, former congressman Bruce Morrison, said that to do so was 'really short sighted' and that the 'ongoing debate is only resurrecting old animosities and stereotypes'. He acknowledged that the United States must 'await the outcome of the current Brexit negotiations' but said they believed 'it should and must make every effort to secure and maintain the GFA which it did so much to bring about'.

The Ad Hoc Committee includes two former US senators, five former US ambassadors and leaders of prominent Irish American organisations such as the American Ireland Fund and the Ancient Order of Hibernians (AOH). Those signing the letter included the most recent US ambassador to Ireland, Kevin O'Malley; Brian O'Dwyer, the grand marshal of the 2019 NYC St Patrick's Day parade; John Fitzpatrick, chair of the American Ireland Fund; former US senator Gary Hart, who acted as the special representative for Northern Ireland for then US secretary of state John Kerry; former lieutenant governor of Maryland Kathleen Kennedy Townsend; former US senator Chris Dodd; former governor of Virginia Terry McAuliffe; former governor of Maryland Martin O'Malley; and foreign policy experts Nancy Soderberg and Jake Sullivan. Soderberg was deputy national security adviser to President Bill Clinton during the critical years when the Good Friday Agreement was being negotiated.

It has amused and interested the US congresspeople to see how their interventions in Brexit and the post-Brexit world have caught fire in the British media and the wider British establishment. Congressman Brendan Boyle said he never did as many international media interviews in one go as he did in the hours and days after tabling his resolution on the floor of the House. He was surprised by looking at some of the social media commentary too (not generally an advisable thing to do).

'In the UK, a person commenting, "Oh, it figures some guy named Brendan Boyle would be involved helping Ireland." I got all sorts of anti-Irish comments. And others were saying "Oh, he's just doing it to win an election, to curry votes," and I thought "Wow, is this person really stuck in the 1880s?" I mean, the idea that there's some big domestic political benefit is laughable. And it just shows a real ignorance about the state of US politics today.'

Congressman King sits on the Financial Services Committee. In April 2019 then British ambassador Kim Darroch met Republican members on the committee, ahead of yet another impending but ultimately toothless Brexit deadline, to discuss financial markets once the UK leaves the EU. King says the issue of the border between the UK and the Republic of Ireland, or the UK and the EU, on the island of Ireland was 'not a key issue' in these discussions. 'They were talking more about derivatives and hedge funds and the financial aspects, but I brought it up. I made the statement to highlight that a number of us will block any trade agreement between the US and Britain if the hard border is reinstated.'

This is something that British diplomats are concerned about and it's understood they have upped the number of contacts between British officials at the Washington DC embassy and powerful US politicians, as the lobbying efforts go into overdrive. One diplomat has been assigned solely to Brexit duties on Capitol

Hill. The statements and interviews that Boyle, Neal and King, among others, have made are fuelling the fear factor. And that is exactly what they are intended to do.

King explains. 'That's a way of using US leverage. One thing I've found is that, in many ways, the biggest ally of Irish America during the Troubles was the British media, because they totally exaggerated our importance. I mean, if we put out some kind of statement in Congress, which maybe about ten of the congressmen even looked at, it could end up being headlines in the Irish and British papers. So that gave us actually more leverage than we deserve. We played it as best we could.'

But during the lead-up to the Good Friday Agreement, the bipartisan Irish American politicians agitating for peace in Northern Ireland had the support of the president. For King, a Republican, that was a Democratic president whom he had not supported, and who, in fact, had beaten his preferred candidate, the sitting President George H. W. Bush.

King says the circumstances are different now. 'First of all, President Trump basically supported Brexit. But also, I don't think he is going to look upon Ireland as the key factor there. I mean, if it's important to the Brits, it will probably be somewhat important to him if he can find a way to work it out. I think that was the thrust of his conversation with the taoiseach this St Patrick's Day [2019]. I mean, the US, I think, is available to help. The State Department is available to help. But he would not show the same interest that Bill Clinton did.'

But just how sharp are the collective teeth of these Irish Americans who are threatening a US–UK trade deal in the face of a 'bad' Brexit? It is incredibly difficult to get a trade deal through Congress – just ask Barack Obama's team who painstakingly negotiated the Trans Pacific Partnership only to have it fall in Congress and be wiped off the table by the next president, Donald Trump. The same goes for the Transatlantic Trade and Investment

Partnership between the US and Europe. There were multiple rounds of negotiations, but it never got close to the congressional level and, again, one presidential term ended, another began, and that was that. So is there real power in the threats?

Pete King continues: 'I want people to think there is. Any kind of trade deal is tough to get through and if they realise that we could generate enough concerned votes in the House to make or break a trade deal.'

Congressman Richie Neal is more circumspect regarding whether there is a role for US politicians, or the US more generally, in Irish affairs. 'People forget that the Friends of Ireland was really born of the Troubles. The idea was to try to move it away from the hard men and the gun-running in the United States. That was the whole notion and there were a couple of groups when they passed the hats and the canisters and all that you knew where the money was going. Certainly Massachusetts was a hotbed of that activity.'

Like many, Congressman Neal feels that granting the visa to Gerry Adams in 1995 was a pivotal issue. But the risk he took in urging it, along with the late Senator Ted Kennedy, was not without political cost for him domestically.

Now Neal is one of the most senior Democratic Congressman in the US House of Representatives, but back then his position was not so secure, and he, like his colleagues, was jostling with the party apparatchiks to get seats on committees – especially powerful committees like the Ways and Means Committee of which he is currently the chair. From 1989 to 1995 Tom Foley of Washington State was Speaker of the House, and the most powerful Democrat. He helped the younger Richie Neal get on to the Ways and Means Committee, and although he was also a proud Irish American, he had completely different views on the granting of Gerry Adams' visa and his younger party colleague's role in promoting it. And not only promoting it, but actually welcoming Gerry Adams to the US and bringing him to his hometown in Springfield,

Massachusetts. A younger colleague he had brought with him on trips to Ireland and had even gifted him a tricolour (which is on display in Congressman Neal's office to this day). 'He was furious with me! He couldn't see past the issue of Sinn Féin and the IRA.'

Speaker Foley and Congressman Neal were together on one particular congressional delegation trip to Ireland in 1989. It was similar to the trip made 30 years later by another congressional delegation that Richie Neal was on, this time led by Speaker Nancy Pelosi in April 2019.

Neal tells the story of how they were crossing the border from Donegal to Derry on a bus. He remembers the British soldiers, what he describes as a 'militarised state' and says he wonders if the so-called Brexiteers know what the situation was like. 'The bus was stopped. British soldiers mounted the bus in full night vision, armaments and searched the bus from top to bottom. A bus of congressmen! The sandbags, the packaging, the border crossing with the checkpoints. The watch towers everywhere. Your conversations were monitored. If you went to the Falls Road and if you went to the Shankill, one of the things that was striking was that you were monitored because helicopters pulled in quickly to watch you. The street sweepers came, the RUC became available immediately.'

He finds the difference now, post-Good Friday Agreement, remarkable. 'Two years ago I crossed the border from Donegal to Derry and my phone pinged – that was it. That's what you're trying to drive home to people as to what happened. It's seismic in its shift.'

The issue of the peace process and the border weighs heavily on the minds of Neal, King and Boyle in particular. Boyle because his father is from Donegal and knows first-hand what life was like. Neal and King because they played such an active role in pushing along the peace process and the Good Friday Agreement and they understand the fragility of the situation in many ways.

Around 175 embassies are based in Washington DC. It is the seat of power of one of the most powerful nations on earth, so almost every country in the world has representation there. Lots of embassies means lots of ambassadors, lots of parties, dinners and receptions. The Irish contingent are frequent guests of the British embassy and vice versa, and relations nowadays are traditionally convivial and pleasant – but that hasn't always been the way.

Congressman Richie Neal remembers a time when relations were quite strained. It was in the lead-up to the Good Friday Agreement when Tony Blair was the prime minister. The Friends of Ireland congressional members were often invited to the British embassy for various receptions. 'We were there for dinners from time to time and it would be awful. I mean, it would be a really nasty get-together.'

John Kerr, a current independent member of the UK House of Lords, was the British ambassador to the United States at the time, a post he held from 1995 to 1997.

Richie Neal said he remembers one night when he was highlighting the issue of the border at a dinner at the British ambassador's residence, and the tone of the evening could be described as far from diplomatic or parliamentary. But, he said, there was still an imperative to try to understand each other's positions, something that may not be there currently.

'On this particular night, the ambassador, John Kerr, lit a cigarette, sipped a drink and said, "Look, I know there are strong feelings in this room, but let's have a go at it." That's how bad it was in those years. The ambassador had to intervene to get us to talk.'

Again, discussion of the Good Friday Agreement comes back to Bill Clinton. Neal says that Blair knew it was the Irish Americans who had convinced Bill Clinton to take the risk of getting involved, of granting Gerry Adams the visa, of sending Senator George Mitchell over.

'We can never get past our own State Department here,' says

Richie Neal. 'They were so pro-British. So that's why Clinton said, "Let's try something else." And his national security adviser, Tony Lake, after a stormy meeting on one particular occasion, made a promise to elevate Ireland to the same status as the Middle East inside the White House.'

The circumstances are different now. The issue with a post-Brexit reality on the island of Ireland is not the same as the situation in the Middle East. The bargaining chip this time is a US–UK trade deal, so what exactly does Congressman Neal intend to do?

During her trip to Ireland in April 2019, leading a congressional delegation, Speaker of the House Nancy Pelosi was resolute in the remarks she made at every speaking occasion. In her address to Dáil Éireann, she put it bluntly. 'Let me be clear, if the Brexit deal undermines the Good Friday Accord, there will be no chance of a US–UK trade agreement.'

Although Ms Pelosi is Speaker, Congressman Neal is a step closer to the front line on this issue in his role as chairman of the Ways and Means Committee.

'My position would be, first, a trade agreement with the United Kingdom is desirable. We would want that. In the next breath, I don't think we're going to take a position that would be any different than the European Union took on the border, and the European Commission. There's been no softening by the other 27 nations in the EU towards the UK and the border. Nobody's position has weakened in the past three years. In the EU, they've all been steadfast.'

But like many others, Congressman Neal is worried about keeping the influence alive into the future. He concedes that the population and interest level of the Friends of Ireland is down from those 'hardest days', it's really just himself, aged 70, and Peter King, aged 75, who remember how it was. And some of the newer arrivals in this current Congress have Irish names and Irish backgrounds but have been slower to join up with the Friends of Ireland caucus or get involved in Irish issues.

Brexit has, in many ways, provided a new rallying cause.

THE DIPLOMATS

EVERYONE'S AN AMBASSADOR

I rish ambassadors understandably possess a vast collection of green ties, jackets and dresses, built up over the years and rolled out for St Patrick's Day events and other Irish engagements through the year.

Anne Anderson was Ireland's first female ambassador to the United States, serving from 2013 to 2017.

She says she had understood and anticipated the level of access that Ireland enjoyed, but only saw how privileged a position the country had on her arrival. It was outlined starkly when she went to present her credentials to the then president Barack Obama in 2013.

'I was wearing a green dress, as one does, especially being the first woman ambassador, and he commented on that. There was a relaxation and a real warmth about the exchange from the very beginning because President Obama is a very courteous, very charming person, but he's also reserved and slightly aloof, but that reserve didn't operate when it came to Ireland. He spoke immediately about the visit they had paid [to Ireland] and his hope was that when he was out of the White House and the Oval Office, when he would have more time and more leisure, he would go back to Ireland and make sure that his kids retain a connection to Ireland. We spoke business – it wasn't all sentimentality – but particularly for the slightly reserved person that President Obama is, it was unusual to see that degree of spontaneity and warmth.'

Comparing notes with other new ambassadors after the event, it became clear that the president had granted a much longer audience to Ireland than he had to the representatives from other countries. 'It was clear from the very beginning that this was a relationship that was working smoothly and well.'

Her successor in the post, the current Irish ambassador to the US, Dan Mulhall, presented his credentials to Obama's successor, President Donald J. Trump. As a diplomat might, he made a point of playing at the Trump International Golf Course before first meeting the president. Trump was understandably delighted to talk about the course during Mulhall's first visit to the Oval Office.

Of course, for diplomats, the reach of Irish power, of that influence, stretches far beyond the White House.

'It's there for governors and mayors and state senates and all of that,' says former ambassador Anderson. 'It gets the doors open for you, but it doesn't deliver for you. You have to walk through those doors and make the pitch, and make the case, and try and achieve the outcomes. These are very busy people at all levels. Time is a very precious commodity for them. They're not interested in spending time on courtesy visits from people, even ambassadors, if they don't feel a sense of connection. So there is no question that being the representative of Ireland, being the Irish ambassador means you have access, means you have open doors, but as I say it's up to us then to go in and to engage the people and not to presume that an open door simply means that you will get a yes to whatever proposition that you are putting to the person in question.

'Time is the most precious commodity and my colleagues, ambassadors from similar sized states, and even larger states, they genuinely sometimes marvelled at the level of access that I had because it was simply not feasible for them to get meetings with people at the level that I would get them.'

In Washington DC, this ability to have a door opened easily is everything. Anne Anderson said the value of the asset cannot be overplayed in a town that is 'powerful and very conscious of influence and hierarchy'.

Her successor Dan Mulhall said that when he arrived in August 2017, he was half-expecting to find evidence of a decline in Irish American sentiment, but he hasn't seen that at all, and in fact has seen quite the opposite.

'Everywhere I've been, I've been pleasantly surprised to find substantial pockets of Irish interest. In other words, people who despite their distance from Ireland – two, three, four generations or more – continue to have some kind of affinity for the country, affection for the country, admiration for the country.'

He echoes the thoughts of his predecessor in that the relationships that he and his diplomatic colleagues have forged need to be cultivated and fostered and cannot be taken for granted. He gets a sense that people feel there is a worth to any meeting they may have with Irish people and Irish representatives.

'I want to make it quite clear that this is not an easy street. It has to be worked for. It has to be minded very, very assiduously by our politicians and our diplomats here to make sure this feeling doesn't wane, doesn't wither on the vine, which it could easily do.'

Ambassador Mulhall says he is constantly meeting people within the administration who have what he describes as 'an instinctively positive attitude towards Ireland, because of their Irish American background', even though that background might have originated generations previously.

When Barack Obama came into office in January 2009, he first appointed Dan Rooney as the United States' thirty-third ambassador to Ireland. The late Rooney served in that capacity until he resigned and left in December 2012, returning to actively perform the duties of chairman of the NFL franchise. He passed away in 2017. As owner of the Pittsburgh Steelers, pro-football hall-of-famer and founding member of the Ireland Funds, upon being appointed ambassador, Rooney and his wife, Patricia,

moved to Ireland, taking up home in the official residence in the
Phoenix Park. His 4 July celebrations became legendary when
he hosted American football games there to celebrate America's
Independence Day.

When Barack Obama was elected to a second term in office,
he needed a new ambassador to Ireland. There was a delay in
naming a candidate, and Ireland had no ambassador for almost
two years. In June 2014, Obama named St Louis, Missouri lawyer
Kevin O'Malley. After the Senate confirmation process he was
sworn into office in November 2014 and arrived in Ireland shortly
afterwards. Ambassador O'Malley served in the post for two years,
tendering his resignation, as is customary practice, on the last day
of President Obama's administration in January 2017.

'To me,' says O'Malley, 'the most memorable and the most
meaningful aspect of being the United States ambassador to
Ireland was the welcome that I received every place I went during
my entire term. I never got used to it. I never got accustomed to
the wonderful, marvellous, warm welcome that is bestowed upon
the American ambassador. Maybe I shouldn't say this, but there
were times that it was very emotional, and my eyes would tear up
at some of the receptions that I received from your countrymen
and women. It had nothing to do with me. It just had to do with
the fact that the American ambassador was there, and they wanted
to do something for the American ambassador.'

O'Malley doesn't like to use the term 'Irish lobby'; he prefers
to talk about those who lobby on behalf of Ireland because he
doesn't think there is a cohesive, monolithic group. He says when
people lobby the administration to do with Irish affairs, they don't
always get what they want, but they will get an audience, for many
reasons. One of them, he says, is that Irish affairs 'don't get us into
trouble' in the way that other interest groups might, citing the
National Rifle Association as an example. His point is that it is not

controversial – not presently, anyway – for a politician to back the Irish, or enjoy St Patrick's Day. Other countries with questionable human rights records, divisive political issues or who are engaged in nuclear activities can cause problems for politicians.

The concept of the Irish in America forming a voting bloc is one subject to much discussion. Kevin O'Malley questions whether there ever was a national voting bloc. The case of the mobilisation of Irish Catholic voters in the time of John Fitzgerald Kennedy is oft cited. O'Malley comes from a staunch Democratic family in a one-time Irish ghetto in St Louis, Missouri called Kerry Patch. He remembers his parents and relations wrestling with voting Republican for the first time in their lives in that election. Wrestling over the candidate they loved, because they feared that if things went wrong with JFK in power, there would be a backlash against Irish Catholics. The anti-Irish rhetoric, the 'no Irish need apply' signs were almost gone by the 1960s, and yet, as O'Malley puts it, 'my parents were struggling'.

That split in Irish Americans, veering between the two parties, became an issue again when Ronald Reagan was running for president. What was left of any sort of unity vote in Irish American communities 'dissipated to some extent', says O'Malley. Reagan was, he says, able to convince working people to vote for him in spite of their best economic interests, and he also enticed them to vote on a lot of social issues. 'When issues like abortion became a part of the political forefront in the United States, many Irish Catholic voters drifted away from the Democratic Party into the Republican Party.' Another reason for this move is what he describes as 'unfortunately, the drawbridge mentality'. As he puts it, 'the Irish got rich and forgot where they came from'.

'They are an issue, those people who forgot. They may be four or five generations down the road, and maybe they didn't have grandparents that had Irish accents, and they didn't have pictures of a run-down house in Westport where they lived. They just forgot that their ancestors came to the United States and caught a huge break, and they're reluctant to give that break to anybody else, and that is maddening to me. "I've got mine, and you're not going to get yours," and to me, that is un-American. These people, then, go into mass on Sunday and bless themselves with holy water and want to keep immigrants out of the country at the same time.' It's clear, as someone who has benefited from the hardships borne by four grandparents who left Ireland in search of a better offering, he's exasperated.

Arguably, this 'drawbridge mentality' was seen in some voters with Irish immigrant heritage in the 2016 presidential election and from some candidates in the 2018 midterm elections as well.

'Absolutely. They just call it a wall now.'

O'Malley belongs to that group of a few dozen Irish Americans who have formed an Ad Hoc Committee to protect the Good Friday Agreement. Their first act was to send a letter to then British Prime Minister Theresa May and Taoiseach Leo Varadkar reminding them of their duties under the agreement to protect peace on the island of Ireland and not to return to a hard border.

He says that, despite having lived in Ireland for nearly three years as ambassador, he would be slow to take a more involved role in Irish politics, or to see the US taking a bigger role.

'I would feel very uncomfortable taking a position in any of the referendums that you've recently had. I would feel very uncomfortable about having any positions about who should hold what office. I don't think that's something that America should get involved in, but the general peace and prosperity surrounding a country that we love, I think that we've felt that we needed to speak out on that.'

He is loath to use the term 'knight in shining armour' to describe the United States stepping in to help Ireland in international affairs, because many people would not agree with that label and it is very difficult to live up to. He is more comfortable with the term 'honest broker', but readily admits that there are few places on Earth that would currently consider the United States to be any sort of honest broker. But he thinks when it came to the Troubles and to Northern Ireland, the US did play the role of 'honest broker'.

'It was a proud moment for us. But I don't think we want the Good Friday agreements preserved because we're proud of them. We want them preserved because they made peace, and they've made prosperity, and they've made life better for the people who live on the island of Ireland.'

———

That strength of Irish representation and diplomatic influence in US political circles goes back to the earliest days of Irish foreign policy. The fledgling Free State of the Republic of Ireland was the first British dominion to be granted representation in the US, ahead of Canada, South Africa, New Zealand and others, when, in October 1924, Timothy Smiddy became Free State Minister Plenipotentiary to the US.

Previously, a presence in the US had been of utmost importance to Sinn Féin when organising the first Dáil Éireann and creating a diplomatic service to argue for Irish independence around the world. Harry Boland was the official representative in the US from May 1920 to January 1922. He and the American Association for the Recognition of the Irish Republic, established by Éamon de Valera, oversaw the campaign for recognition in the US. America was seen as key to the push for independence. The recognition that was granted in 1924 prompted Joseph Walshe in the Department of External Affairs in Dublin to remark that America was the only country 'with which our relations are entirely free and independent from any outside control'.[14]

The diplomatic ties between the two nations operate at a level that is unique. No other country has quite the same relationship with Ireland as the US does. The official residence of the US ambassador to Ireland is in the Phoenix Park, across the roundabout from the official residence of the president of Ireland. It is a house owned by the Irish government, rented from the Irish state for next to nothing by the Americans. The sitting US ambassador is afforded cheaper and more luxurious accommodation than the taoiseach of the day. This is despite the fact that American ambassadors tend to be wealthy political appointees, in contrast to Irish ambassadors, who are civil servants and career diplomats.

With the US recognising the Irish Free State in 1924, the newly independent Ireland went on to open not just an embassy in Washington DC, but several consulates in the 1920s and 1930s including New York, Boston, San Francisco and Chicago. More have opened in the intervening years and, with a consulate in Miami set to open in 2020, the Irish diplomatic network in the US is as big as that of the UK.

—

It is inevitable that when one party is in power for a period of time, the connections with those in that party are stronger than in the other party, and this is something that recent rounds of Irish diplomats in the US are always conscious of, particularly outside Washington where mayors and governors can be in office for lengthy

14 NAI, D/J, Letter Books, 1922–25, Walshe to Secretary, Department of Justice, 21 October 1924. As quoted in 'Recognition of the Irish Free State, 1924: The Diplomatic Context to the Appointment of Timothy Smiddy as the First Irish Minister to the U.S.' Paper by Professor Bernadette Whelan, University of Limerick and Associate Fellow, Department of History, University of Warwick, UK. October 2014

periods. Given a more traditional affinity with Presidents Clinton and Obama, the election of Donald Trump could potentially have posed an access problem for Ireland. But diplomats must always have an eye on context when they are cultivating potentially powerful relationships. One of the most valuable relationships in recent times has come in the shape of the president's Chief of Staff Mick Mulvaney,[15] who is a great friend of the Irish embassy. The relationship began when he was a Republican congressman, quite active in the Tea Party branch of the GOP, and was approached by the Irish embassy team to discuss immigration.

Anne Anderson says there was a deliberate effort in recent times to develop contacts on the Republican side of the House, and she became quite friendly with Mulvaney. In fact, the congressman was at a social gathering with members of the Irish diplomatic team when he received the call from the Trump transition team, in the aftermath of the November 2016 election, offering him the role as Director of the Office of Management and Budget. Little did anyone know that he would end up as Chief of Staff, the most crucial role in any administration, as that person is the gatekeeper to the president.

Current ambassador Dan Mulhall says the embassy team makes every effort to cultivate people 'on both sides of the House', and does it 'quite assiduously'. He admits there is a challenge there in that, as he sees it, there are probably more members of Congress who are Democrats who feel a strong connection to Ireland, and

---

15 Technically, at the time of writing, Mick Mulvaney is Acting Chief of Staff, and he is also still the Director of the Office of Management and Budget. However, a number of individuals in the Trump administration hold positions in an 'acting' capacity even though they have been appointed by the president to that role. The reason is that many cabinet and presidential appointments need Senate confirmation to become official and so the individual can swear an oath of office. Given that the margin of Republican control in the Senate is so narrow, and the politics are so polarised, it is difficult for any Trump appointee to get approved. By keeping them in 'acting' roles he avoids that. Although in the case of Mulvaney, he was already approved by the Senate for his first appointment as OMB Director. Nevertheless, for the purposes of description here, he is referred to as the Chief of Staff as that is what the president calls him, and others refer to him as, even though he is technically still 'acting' at the time of going to print.

who feel that connection is beneficial to them politically. So while the powerful Friends of Ireland Caucus on the Hill may be slightly weighted in favour of Democrats currently, he says there is still 'a bunch of Republicans who have an Irish connection', pointing to Vice President Mike Pence as the most senior.

Ireland has had a run of recent good luck with presidential chiefs of staff. In the Obama White House, the role was held by Bill Daley from Chicago and, in later years, by Denis McDonough, whose cousins still live in Galway. Trump's first chief of staff was Reince Priebus, who did not have any connection to Ireland, but he was replaced midway during the President's first year by John Kelly, and he in turn was replaced by Mick Mulvaney. A lot of Irishmen bearing influence with the holder of arguably the most powerful office in the world.

Anne Anderson says that you have to be careful of guarding such powerful relationships and not overusing them, regardless of the pressure you might face. 'It's not at all that you're calling them every day or anything like that, you have to be very respectful of their time. But if you want good relationships with them, they will know that you will only call them when you really need them and therefore they will be responsive. So it's a great asset but one you use very sparingly. And it's all the more effective because they know you use it sparingly.'

This power nexus is intoxicating in a town like Washington DC. When others know that you're connected like this, it makes them more responsive to you by virtue of your connection to that power.

Ireland is a relatively small country, but is incredibly influential on Capitol Hill, incredibly influential with presidential administrations, and has senior and powerful connections in corporate America. This is all despite the fact there's not necessarily a domestic payoff for the folks who are involved, but it is because of some other deeper connection.

THE PRESIDENT

IT'S WHO YOU KNOW ...

H e looks at it from time to time. An old glass bottle filled with soil and dirt. It was taken from the last known building that the Cassidys lived in – an old farmhouse outside Rosslea in County Fermanagh, where the ancestors of Bill Clinton's mother hailed from.

'I'm not alone in having that bottle full of dirt,' says the former president wistfully. 'I think that most of us who have any kind of Irish roots are proud of the writers and the poets and the musicians and the politicians and the dramas, and we can normally agree on that, even when we're deeply divided at home.'

In total, 22 of the 45 US presidents have claimed Irish heritage.[16] The eighteenth president of the United States, Ulysses S. Grant, was the first US president to visit Ireland, although he came after his term ended. His great-grandfather came from County Tyrone. He delivered a speech outside Dublin's City Hall in 1879 remarking on how many more Irishmen there were in the United States than in the whole of Ireland. The United States' seventh president, Andrew Jackson, had the closest ties to Ireland. Both of his parents were born in County Antrim, and he was born in the United States just two years after they emigrated.

Bill Clinton admits that he did not know much about Anglo– Irish relations and history before he went to study at Oxford University, at around the same time as the euphemistically named

16 According to EPIC, the Irish Emigration Museum. <https://epicchq.com/us-presidents-with-irish-heritage/>.

'Troubles' flared up. As a student, Clinton went on a trip to Dublin and then started following Bernadette Devlin's election campaign. He knew that at least part of his family – the Cassidys on his mother's side – had originally come from Northern Ireland, and when he saw the tragic nature of the situation, he took an interest.

Fast forward through his years in Arkansas to 1991/92 and he was running for president. The talk, he says, among Irish Americans in New York at that time was whether the US should be doing more and, he says, 'a lot of them were interested in supporting me'.

A presidential forum was held in the run-up to the New York Democratic primary, organised by Irish American officials and activists. President Clinton says his good friend, then congressman Bruce Morrison, who had gone to law school with himself and Hillary, was involved. Morrison wanted Clinton to take part and so he did. Clinton points to this forum as the beginning of his friendship with Irish American organisers in New York like Niall O'Dowd, Brian O'Dwyer and his father, Paul O'Dwyer.

Publisher of *Irish Central*, living in the US since 1979, Niall O'Dowd says the timing was perfect for Irish issues when Clinton was trying to win the New York primary. The Irish lobby offered a cause that he could get behind.

'Back in the days of Vietnam, you'd never have the Irish getting a hearing at the White House,' says O'Dowd. 'But take the period when Clinton was in – there were really no issues. The Cold War was over. There was Bosnia, but there wasn't a huge, international al-Qaeda-type issue. So when you had an activist Irish group come in, we got Bill Clinton's attention.'

There was a ready-made nucleus of people who had been working together on the immigration issue since the 1980s. He cites congressmen Morrison and Donnelly and their efforts to get visas for Irish people. 'We came together as a very strong lobby. We got to know people like Ted Kennedy, who was pretty good to us. Chris Dodd. People who became critical in these issues.'

O'Dowd says the Irish community didn't necessarily have huge numbers, but it was the class of person that they knew. 'Who do you know? How can you use them? How can you leverage them? We leveraged successfully in immigration. We got 40 thousand Donnelly visas and 40 thousand Morrison visas, roughly. And then we leveraged very successfully in the peace process.'

Former mayor of Boston Ray Flynn was involved in that forum in New York City in 1992. He was on the panel questioning the candidates.

'The question I had was, if you're elected president of the United States, would you appoint a peace envoy to Northern Ireland? I was putting them really on the spot! A number of them didn't want to commit to that because all the press in America at the time was not for helping Ireland. They were mostly helping Great Britain. So I asked the question and Governor Clinton said, yes, he would appoint a peace envoy if he was elected president. Well, that caught on all over the country. It was in all the newspapers and it was the most significant statement ever made on the peace movement in Ireland. But the British government was not happy with it at all – the British prime minister even told me that later.'

Bill Clinton recalls that agreement to appoint a special envoy, before he was elected, and knows it was a risk. 'I knew that it was a controversial thing to do, but I knew that America had an unusual impact or potential impact on the situation, not only because so many people gave money in America to Irish charities, and some organisations contributed money to the IRA. But I just thought we had the largest Irish diaspora in the world and we ought to get involved.'

Former mayor Flynn says that big Irish diaspora responded in kind to the then candidate's promise. 'The next thing you know, the Irish Americans were organising these social clubs, civic clubs – San Francisco all the way to Portland, Maine to Boston, New York, Chicago, Philadelphia, all over the country, mostly in these traditional swing Democratic cities with a large Irish American population. They're the ones that elected Bill Clinton as president. The Democrats were solid behind Clinton. The Irish Americans were solid behind Clinton.'

This was a campaign promise and Clinton acknowledges that he could have backed out of it, as politicians often do. What is promised on the doorsteps and what materialises are often not the same.

'Well at first, in fairness, some of the more ardent supporters of greater American involvement thought I wasn't going to follow through because I got a lot of grief about it, including from some of the parties, like the then speaker of the House [Tom Foley], who was an Irish American Protestant from Washington State, and even some of the people like Senator Kennedy. Some of them were afraid I would mess with the special relationship with the US and the UK.'

Clinton says he also got a lot of negative media attention in the UK about it, but nevertheless, he followed through and appointed Senator George Mitchell to be US special envoy to Northern Ireland.

'After the '94 elections Senator Mitchell was leaving the Senate and he had declined to be considered for the US Supreme Court. I thought he would have been a magnificent Supreme Court judge, but he didn't want to do it and he had a new life to start, so I got him involved in this. I told him it was a part-time job. Which turned out to be only partly true. He's been telling jokes about "my part-time job" for I don't know how long. He says, "I realise now what he meant by part-time job is he would allow me a few hours a day to sleep."' Clinton chuckles to himself, knowing full well that, at the time, neither man anticipated it would only be a part-time job.

Clinton also had a number of Irish Americans in his entourage and in his administration, and they feel their internal pressure helped him keep the focus too. Adviser and still close friend Terry

McAuliffe, who went on to be governor of Virginia and toyed with running for the White House in 2020 but ceded that path to 'another proud Irish guy', Joe Biden, says they all brought pressure to bear too. 'He was surrounded by Irish. We were everywhere. He had the most African Americans in his cabinet and his administration, but he had the most Irish, too. And he was a fun-loving guy, so we all got to spend a lot of time with him.'

McAuliffe thinks that there were three Irish presidents – Jack Kennedy, Ronald Reagan and Bill Clinton.

While he had special feelings for Ireland, Clinton says that he identified three situations in the world that he wanted to try to change. The situation in Northern Ireland, the situation in the Balkans and the ongoing conflict between Greece and Turkey over Cyprus.

'I always thought that the Irish themselves would solve this if we could just get everyone talking. That's why I did it. It had nothing to do with anything else. I thought it was the right thing to do.'

Appointing an envoy was a critical move, but before that, he says the initial pivotal decision was whether to approve Sinn Féin's Gerry Adams for a visa to come to the US. 'It was all about whether he promised at the time not to raise money in America, not to raise all these issues about whether it was an IRA front rather than a political mission. But we worked through that and I decided to do it.'

He says he had the support at the time of the National Security Council but he faced the 'intense opposition' of the State Department, 'including not only the secretary of state, but the American ambassador to the United Kingdom, Admiral Crowe, who was a very close friend of mine and former chairman of the Joint Chiefs of Staff.

'They were afraid I was doing irreparable damage to our long-term relationship, but I knew that couldn't be true. I knew we had too much in common and our interests were too intertwined for that to happen, and I just felt like somebody had to do something to break the logjam. I was hoping that getting Gerry Adams a visa would do it and, if you remember, it took a little while before the ceasefire was declared, and during that little while I took no little heat for having done something that didn't produce immediate results.'

The visit he made to Ireland, North and south, in 1995 was also viewed as instrumental. A cross-party delegation from Congress came with him. One of those was Republican Congressman Jim Walsh from upstate New York. He recalls what they experienced and how it was relayed on US television as a turning point in the engagement and in the profile the issue received domestically in the US.

'I saw the injustice, I went there, I understood it. I was with Clinton at the Mackie's plant. I'll never forget what really solidified it for me. There were doors for the nationalists at one end of the factory and there were doors for the unionists at the other end, the loyalists at the other end. They would work together and then they would leave by separate doors. And what really made it strike home for me was a big ceremony where they had a couple of children who lost a parent in the Troubles and other political leaders speak. Then Clinton spoke, and in his speech, he said that it was time to forgive and forget, and after he'd said that, way in the back, I heard "never". As loud as the president's speech with a microphone was this "never" that echoed through the whole place and I said, "Holy shit, that tells me a lot." Never. That's a big word. And there was a lot of that talk then. Never, it will never be resolved. It will never, ever end. There will never be justice. And I just said to myself, "Yeah, there's going to be justice someday. And I want to be part of it."'

Clinton says he became particularly concerned about the path he was pushing for when the Omagh bombing happened on 15 August 1998, killing 29 people, including a woman pregnant with twins. 'I was so worried about it. You'll remember I raced over there again.'

Indeed they did. He and First Lady Hillary Clinton arrived in Omagh just a couple of weeks after the bombing, on 3 September. Prime Minister Tony Blair and his wife Cherie were also there that day. 'We all worked so hard to keep the lid on the process when something bad like that happened.'

However, he thinks that perhaps the most important thing he did was to change the way St Patrick's Day was celebrated in the White House, elevating it from just a social call with shamrock to a bilateral meeting between the taoiseach and the president and a reception.

'What we did was, we had a dinner and an event at night with lots of music and singing and collaborating, but it enabled me to give all the parties a chance to come to the White House and meet with them all, often while everybody else was partying. But it was informal, relaxed and consistent and it gave me a chance to reassure them all that we were involved, and then, of course, all year long our security staff was working on it.'

Throughout the run-up to the Good Friday Agreement, President Clinton says he needed all sides to know that the United States was not going to walk away easily. 'I sent a signal to people on both sides that America was all in, but not trying to dictate the final shape of the settlement – just there had to be one and it was crazy to continue with this division and the consequences of it.'

It could be easy, and perhaps somewhat cynical, to say that the first promises Clinton made to that group of Irish Americans at the presidential forum in New York was all about winning the New York Democratic primary. There may not be an Irish voting bloc per se, but in New York there are a lot of Irish voters. And he needed the voters in the primary in 1992, then the general election, and then the re-election campaign in 1996 – at which point his involvement in Northern Ireland may not have been beneficial for him with non-Irish American voters.

'I didn't think it was a bad thing, because I had previously thought that the vast majority of the Irish in America who had a consciousness of their Irishness didn't think of themselves first as Irish Americans. But millions do. So did I think it would be good politics? Probably, but only if they trusted me to be an honest broker because the vast majority of Irish Americans didn't approve of the violence of the IRA or the Unionist extremists. They didn't want any of that. They wanted their homeland to be peacefully, if not reunited, at least tied together, cooperating, working together. They wanted it to function in a way that would allow Northern Ireland to raise its living standards to the level of the Republic. And within Northern Ireland, for the then Catholic minority to raise its living standards and the range of its civil rights to the level of the then Protestant majority.'

His former co-chairman of Irish Americans for Clinton and former mayor of Boston, Ray Flynn, thinks it was entirely a good political decision. He says Bill Clinton going 'all in' for Northern Ireland changed everything, and it was viewed as a victory for the Irish American community all over the country.

'As a result of that he won the election, of course, and he and the Irish were very, very strong and they took pride in their accomplishment. I went all over the country many, many times. And it was just a sensational time to be Irish American in America because of the pride.'

But that's not the case any more, says Flynn. 'Right after the Clinton era, it kind of fizzled out. Some people still try to keep it going, but most of the politics changed dramatically. There wasn't the political benefit of being deeply active in organising Irish groups in the country.'

So while Bill Clinton got the benefit of an Irish American vote that could be organised and engaged based on his involvement in the peace process, he too sees that, nearly 25 years later, it is clearly not the same any more. The cohesion is gone.

'There are clearly Irish Americans who support the anti-immigrant, build-the-wall, anti-trade politics of the current government, but there are a lot who don't. And there are people who believe that the most important thing is to keep what they have now and others who believe that, as Irish Americans, they owe other people the same chance they had to do well in America.'

However, Clinton thinks Irish Americans could be re-energised and a substantial cohort of voters could be mobilised once more to help get somebody elected – but he thinks that now exists in both parties.

'There's a partisan split as there was at the beginning of the Civil War, when the Irish were divided over whether it was a good or bad thing to integrate African Americans into the economy of places like New York. I think there's still a lot of people who believe that it's very, very important to support peace in Ireland and to support the development of Northern Ireland.'

While Clinton does not think there is an Irish voting bloc now, he thinks there could be one at some point – but with limitations. 'Do I think it is likely we'll run anybody that will be able to get the percentage I got when the Irish peace process was in the headlines all the time? Probably not.'

But he says that, while he got a sizeable vote from Irish Americans when running for election, he did not get them all. 'Even when I ran, there were a lot of people who loved what I did in Ireland, but they were still loyal enough to the Republican Party and agreed enough with their policies that they didn't vote for me.'

And that's just how politics should be, according to the former president.

'It'll never be a monolithic vote and it probably shouldn't be. People should vote with their consciences, but the power of the culture can tug tightly or loosely at election time depending on what the most important issues are at that moment.'

Whether or not there are votes in it for American politicians, he does think there is a role for the US to continue to play in Irish affairs, particularly given the uncertainty created by Brexit. Whether it happens or not, whether it's hard or soft, long or short in duration, there will be consequences for Ireland. Clinton says the US should be involved on both sides of the border.

'I think we ought to be plugging for peace and looking for good things to happen. I think we ought to reinforce our efforts to make economic improvements in Northern Ireland. Things haven't recovered in line with the growth rate of the Republic, but the last time I was there, there were plenty of cranes up in Belfast.'

He laments the Brexit referendum campaign that was run in 2016 and feels that 'a lot of the stuff that was fed to the English and the Welsh was not accurate'. He thinks if they had had accurate information, the outcome might have been different. 'It's very sad to me that it happened, but it did happen and now we just have to find a way.'

At the time we're talking, late in May 2019, there is renewed pressure to put the question of leaving the EU to the people of the UK once again, to hold another referendum. This time the thinking is that the electorate would be armed with more information than they had in June 2016 and would have a realistic sense of what's involved in extricating the UK from the EU.

The former US president is clear on where he stands on that question. 'I wish they'd have another vote and undo it, but if they're not going to, I think it's imperative that the Irish just keep holding their breath, keep talking, keep working together, maintain the status quo, until they see the shape of a final deal that is acceptable to both the European Union and the UK parliament and then we'll see what happens. But there's no question that one of the casualties of the Brexit vote has been continued progress in Northern Ireland.'

In particular, a casualty of Brexit has been the failure to resume the Northern Ireland Assembly, dissolved by the late Deputy First Minister Martin McGuinness of Sinn Féin in January 2016. The focus of the UK government on Brexit, and the kingmaker role played by the MPs of the Democratic Unionist Party in holding up Theresa May's (now Boris Johnson's) coalition government, has meant that getting the institutions back up and running has been beyond the grasp of Northern Ireland politicians.

The blame has been laid at the feet of the leaders of the two main parties: Arlene Foster of the DUP and Mary Lou McDonald of Sinn Féin. President Clinton does not agree with that view.

'How could they be expected to make a deal to work together not knowing what it is they're working on or even the entity they're working in?' he says. 'So we all just have to take a deep breath, keep talking and keep trying to develop Northern Ireland in a way that's fair to both communities and then events will unfold in a way that will tell us what we can and should do, I think. It's almost impossible to conceive of any government other than a unity government for Northern Ireland. It's almost impossible for them to agree on the terms of their unity not knowing what the shape of Brexit will be in the end.' An acknowledgement from the former president that the concept of a United Ireland is now more than just a talking point. It could be a realistic prospect.

'Northern Ireland needs to stay in the UK because they're not quite ready to join the Republic, but this Brexit thing really threw a monkey wrench into the Irish peace process. With this young woman journalist, Lyra McKee, who was killed and all the violence that has happened recently, we've got to concentrate on telling people that you don't want to give up the benefits of peace. We'll have to figure out once we know what the final shape of Brexit is whether the people of Northern Ireland want to stay or go.'

He thinks if the Scottish people knew what would have happened with Brexit, they probably would have voted for independence in the vote to leave the United Kingdom in 2014.

'I don't agree with the Brexit vote. I think it was a terrible mistake.' He went to Ireland three times in a twelve-month period in 2017 and 2018 to, as he puts it, 'try to minimise the consequences of Brexit and all the uncertainties. I've worked closely with the Irish but also with the British government and I told everybody to relax.'

The former president says he's not currently involved in a formal way in what the Trump administration may be doing, or may plan to do, but as far as he's concerned, he stands ready and willing to serve the Irish people if he can.

'I've continued to stay in touch with the British representative for Northern Ireland and, when appropriate, with the leadership in the UK. When I'm in Ireland, I try to check in with the government there. When I'm in Northern Ireland, I try to check in with the leaders there, and I still stay in touch with all of them as I can.'

And as if to demonstrate the level of influence Ireland can still wield at this point in his life, he adds: 'I want them to call on me if they think they need me, and I try to stay out of everybody's way if I'm not needed.'

—

The greening of US politics reached its heyday in 1960 with the election of an Irish Catholic, John Fitzgerald Kennedy, to the office of president of the United States. 'He would never have been president had he not been Irish,' his widow Jacqueline Kennedy wrote in a letter to the president of Ireland on 22 January 1964, thanking him and the people of Ireland for their condolences following the assassination of the president. The mobilisation of Irish, and expressly Catholic Irish, voters that occurred in the

1960 presidential election is not a feat that anyone feels could be repeated in the modern political era.

The Kennedy name, the Camelot era, has woven a mystique throughout Irish American politics ever since. Since the assassination of the president, his relatives have held US Senate and congressional seats – the current one being his grand-nephew Joe Kennedy III, aged 38, elected in 2013. He's the US congressman for the 4th district in Massachusetts, which encompasses the western suburbs of Boston to the south of the state.

Named after the elder brother of his grandfather, Bobby Kennedy, who was killed in action during World War II, his father was Joe Kennedy II and he is the third.

Though still relatively young and inexperienced at national politics, this lawyer and former Peace Corps volunteer is one who many in the Democratic Party hope will one day, possibly in 2024, make a run for the office once held by his grand-uncle Jack.

A career in public service was always a natural choice for him. He says elected office was not something that was as foreign to him as it was for most people because of his family connections. His father was in office from the time Joe was six or seven, and he was familiar with the work of his uncle Teddy and other elected family members, and obviously his assassinated grand-uncle JFK and grandfather Robert. 'So for me, it wasn't something that was all that crazy or outlandish to do.'

He didn't go straight into it, though. First, he volunteered with the Peace Corps (a US government development and aid agency) in the Dominican Republic for two and a half years. Then he went on to law school and worked as a prosecutor in Massachusetts. It was during this period that he was inspired to seek office – buoyed up, he says, by seeing how inspirational the United States could be with laws and structures designed to protect and help, but also because of witnessing how the criminal justice system sometimes really doesn't work in the way it is designed to.

He tried to keep people out of the courts system entirely and says his role as a prosecutor in lower level criminal court was 'to try to use the tools that you had to address the underlying circumstances that led that person to be arrested in the first place'. He realised that policymakers had more leverage than prosecutors did and decided it was time for a change of direction.

Much of the US and his native northeast is ravaged by an opioid crisis and this is another situation that led him to seek a new career. 'If you're going to look at something like an opioid epidemic, you have to look at the structures that allowed for opioids to be as prevalent as they are. From the complete lack of mental and behavioural health systems and the lack of addiction resources for people suffering with addiction, to treatment facilities, to accessing care within our prisons and jails – all of that led to criminalisation of people that were sick. And we saw – and are still struggling with – the consequences of that structure. So that helped me come to the conclusion of saying, "Hey, if you really want to influence that further upstream, you've got to get involved with setting the policy."'

And so this Kennedy followed in the family footsteps and went to Washington.

Like many other Irish American politicians, he credits his Irish upbringing with that call to service, that quest for social justice.

'In my family and obviously, particularly, my father's family, our Irish heritage is something that was very important to us and still is. The stories that my Uncle Teddy would tell of my Uncle Jack's trip to Ireland in 1963, still with such pride and such laughter, and still laughing at the same darn lines every single time. And my dad's trips to Ireland, to the south and to the North. The pride that all the members of my family have and continue to hold about their heritage.'

Congressman Kennedy has told his family's story of immigration a few times on the campaign trail, going viral on several occasions, comparing his stories of his family's experience of 'No Irish need apply' with similar tales of discrimination and racism that are occurring in the America of today.

'The significance of their story, of a family who left because of an abject failure of government in the midst of the potato famine, when there was enough food grown to actually feed every person in Ireland, but was taken and exported. To know that the Irish population today is still less than it was before the Famine. And then to understand that migration and the consequences of it for a family like mine that ended up on a coffin ship into Boston. I've had a conversation with the former Speaker, Paul Ryan – his family were on a similar ship into Canada and then across western Canada and south into Wisconsin. Mick Mulvaney, now acting chief of staff, his family ended up in South Carolina. There's obviously an awful lot of political differences, but the stories are shared. For my family, this country has led to enormous opportunity and success for us, but it is a country where our ancestors were not necessarily greeted with open arms. Our family, like many others, had to find its way through that and formed a community around it, that was able to band together and form enough of a political coalition to actually take power and then fight for those that had been less fortunate. Unfortunately, a lot of those same structures and challenges exist for immigrants from other nations today.

'When I think about what it means to be an Irish American, it means sharing a part of that same family story with a sizeable and significant percentage of the American population whose descendants came through some very tough circumstances. And I think it allows us to empathise with those that are going through similar circumstances today.'

The Kennedy story of the sharp rise from impoverished immigrants to the White House in just a couple of generations is the embodiment of the American Dream in many ways, notwithstanding how Congressman Kennedy's ancestors may have gone about it. But young Joe Kennedy says he does not run expressly on a ticket of being Irish or Irish American, though he jokes of his curly red hair and freckled pale skin that 'I'm not so sure anybody that sees me thinks I'm anything else but Irish.'

He may be identifiable as 'the Irish guy' in his district, but his family name introduces his backstory before he ever opens his mouth. His part of Massachusetts is that same district with the highest concentration of people claiming Irish heritage anywhere in the US, but he doesn't think there is what could be called an Irish voting bloc or even a set of issues that could bring Irish America together to vote for a president. He does, however, appreciate that Irish America holds a unique place in the US political calendar.

'One of my favourite days of the year down here [on Capitol Hill] is the Speaker's St Patrick's Day Lunch. That is a unique moment here. We don't do lunches with the Speaker like that for anybody else. Bipartisan delegations don't come together like that for anybody else. It's done to acknowledge that shared heritage that we have, but also to recognise that there's an Irish diaspora that will come down as Democrats and Republicans. It's one of the rare times that unites those factions. The shared heritage is, in many ways, far, far stronger than whatever political challenge that might be confronting us at that moment.'

Congressman Kennedy says he hasn't noticed a swing in his district of Democratic Irish Americans becoming Republicans, but he does think that shift has happened and that it happened before he entered politics.

'I haven't seen a shift, a modern shift, in terms of Democrats to Republicans. I think part of that probably had taken place before I was more active politically. But it is certainly not rare to find Irish Republicans to the extent that it might've been when President Kennedy was in office. Certainly there's a number of Irish Americans that are Republicans – some pretty conservative Republicans. But when Mick Mulvaney was in the House, he was also supportive of immigration reform. Not to say that he'd agree with, necessarily, the bill that I would write, but he was supportive of immigration reform. And Paul Ryan, before he was Speaker, was a supporter of immigration reform. There are differences on issues on taxes or regulation or whatever else, and some of them are bedrock pieces of Democratic and Republican ideology, but there are also common threads coming through. I don't want to say all of us, but I think many of us are strong believers in an active and engaged relationship with the United States and Europe, NATO, EU, and understand that, in a world with increasing challenges, you need to have an increasing number of friends.'

Many Irish American politicians and those who are politically active would say that the mobilisation of Irish Catholics and Irish across the US led to the election of President John F. Kennedy. So can his grand-nephew Joseph Kennedy III see another mobilisation happening that might lead to the election of another President Kennedy?

He laughs, blushes and shuffles in his seat a bit. 'Um, ah, em ...' He exhales and laughs again. 'I thought you were going to say could they elect a President Biden!

'So looking in the immediate term for president – why don't I put it that way?' he says, laughing awkwardly. 'I think potentially. But I think there's also going to be an understanding that an Irish American community here is not unlike a cross-section of the rest of America where heritage still matters. It's also going to be up to our candidates to be able to speak to values and unite different coalitions around a common vision for our country.'

He laughs and shuffles in his seat again. 'For now, though, we have two Irish Catholic folks already in this race. One named Biden and one named O'Rourke. So we'll see how that goes down on the road. But as to anything further, well, we'll see!'

So that's not a 'no' to him running for the White House himself?

'Put it this way, you'll be able to have your baby and come back and cover that race if it happens. [I'm heavily pregnant during this particular interview in his Capitol Hill office.] We'll hold off for you till then.'

I joke that I'll see him on the campaign trail in 2024 so.

'There you go. Please do!'

You heard it here first. #Kennedy2024

GREEN TIES, GREEN FOUNTAINS AND GREEN BEER

'My guess is a valet would lay out a green tie for him because he'd be reminded by somebody. That's it,' says one Barack Obama aide of the 44th president's St Patrick's Day sartorial choices.

The 45th president needed reminding too. Former White House press secretary Sean Spicer tells the story of how he reminded President Trump the night before his first St Patrick's Day in office to make sure he had a green tie to wear. Spicer – a proud Irish American from the tribe out of Rhode Island – wore his own green tie but brought a spare to the office in case the president needed it. And he did.

'The billionaire president wore my green tie that entire day, including to the events with the taoiseach. He must have liked it because I've never seen that tie again,' says Spicer. He doesn't expect to get it back either. 'I don't think I had any illusion that that was ever going to happen,' he jokes. President Trump wore a different tie to the subsequent St Patrick's Day celebrations he's so far presided over, so the whereabouts of said tie are still unknown.

There is universal agreement that St Patrick's Day at the White House (which almost never falls on 17 March due to the US legislative calendar) is the most important day in the Irish–US political calendar. In recent times, the day has started with an early breakfast gathering at the Naval Observatory, the home of the vice president. From there the taoiseach and vice president of the day, and their respective entourages, travel by motorcade to the White House for a bilateral meeting with the president. Then they all travel together to Capitol Hill for a lunch hosted by the sitting Speaker of the US House of Representatives. The leaders then part ways for a few hours before reconvening in the White House in the early evening for the official presentation of the traditional bowl of shamrock. Then there is a rowdy-ish party hosted by the Irish embassy before a more select group (often including vice presidents) gathers at the Irish ambassador's residence for a supper. So all in all, it's a fourteen-hour takeover of Washington DC by the Irish. Quite the power play.

But it wasn't always like that. The gifting of shamrock to mark St Patrick's Day first happened in 1952 when then ambassador John Hearne brought the plant to the White House to give to President Harry S. Truman, but he wasn't there at the time. The ambassador did the same thing in 1953, but this time in the now traditional Waterford Crystal bowl. It was gifted to President Dwight D. Eisenhower, and he, assuming it was on behalf of the Irish people and not just the work of an enterprising diplomat, sent a thank-you note to President Sean T. O'Kelly.

A memo from the time shows this gift wasn't actually instructed to be made on behalf of the president, but the thank-you note was gratefully received and the gift of shamrock then became a tradition. The first taoiseach to make the presentation in person to a US president was John Costello in 1956, when he met Eisenhower. And President O'Kelly himself made the journey to give the shamrock directly to Eisenhower in 1959, the first Irish president to formally visit the United States. And so the gift has been given every year since then. It has not always been made by taoisigh – sometimes it has been by a minister or an ambassador; and it was not always gifted directly to the US president – sometimes it has been a cabinet secretary or other aide.

However, during the peace process and in the run-up to the Good Friday Agreement, it was decided that it should be the taoiseach who came every year to elevate the stature of the call. So in those years, it became not just a shamrock presentation but also a substantive meeting between the taoiseach and the president, and an evening event. This change first occurred while President Bill Clinton was in office. He says he took the decision to lift up the occasion expressly in an effort to push along the Good Friday Agreement.

'I thought, first of all, if we could get them all here to the United States, they would be a little bit out of the firing line at home and then they could talk with each other. And I thought they'd be here, so I and others could talk with them. And I was hoping that, in doing both things, we'd be able to keep everybody together and keep making progress. That's really what we used St Patrick's Day for.'

President Clinton says that while the bilateral meeting was important, it was the evening event, usually a black-tie gala dinner, where the real meaningful discussions and negotiations took place.

'I can't tell you how many times I'd grab somebody by the hand and then we'd sneak upstairs to my office in the White House residence and visit. Sometimes, I'd meet with two or three people at one time, sometimes I'd meet with people alone. But I thought it was obviously the right thing to do.'

He never envisaged that it would grow to the scale it has or that it would be still such a big event on the calendar. 'I think it just grew into that. All I knew was that I wanted to see them all every year, no matter what. I wanted them to know that it wasn't just my administration – that it was me, personally involved. And what can I say, I love the music.'

Long-term Irish Americans will tell you that the evening party was beyond compare during the Clinton years. An Irish hotelier in New York, part of the Irish American Democrats organisation and friend of the Clintons, John Fitzpatrick explains: 'It was a black-tie dinner. We went to a reception and we sat down, and we had dinner in the early days with Albert Reynolds, as taoiseach. The president welcomed us into his home. I remember it got so crazy after a few years, there were just so many – 350 people were there at the dinner. I remember one night, President Clinton, they have their own section where they go into an elevator and they have their private quarters. I remember him saying, "We're going to bed. Just turn off the lights when you leave." Because we, the Irish, took over the White House. We nearly took it for granted, as if it was someone's home and we're just going there for dinner. Then it changed a little bit because Bush got in, and of course he didn't have the relationship with Ireland that the Clintons had. I think that was an exceptional time for us.'

President Clinton himself remembers the great parties. 'Oh, it was amazing. And we loved it. Everybody who could sing and play wanted to come and sing and play because they knew that they were there celebrating the potential of a whole new day in Irish relations. That also made a huge difference. It was great for me. And I think the Irish all felt that way too.'

Former Clinton adviser and governor of Virginia, Terry McAuliffe, says there was nothing like those parties at the White House for scale or fun. 'We actually had to do a tent in the back yard because there was so many Irish. We couldn't fit any more people in the White House. There were thousands of Irish, drinking, singing, having a good time, but he loved it. He loved it. And given that Hillary had grown up in Chicago, knowing the Daley family, the influence of the Irish was something for her as well, so she enjoyed it too.'

The former president adds, with a chuckle, that nothing got in the way of those St Patrick's Day events. 'I was often interrupted by problems around the world on golf courses and other places, but I don't think there was ever a time when I was president that

St Patrick's Day was interrupted by a crisis in another part of the world.'

The black-tie evening gala has ceased, and the evening reception is now in the late afternoon and no longer involves the president freely mingling with the general invitees. There is something of a greeting line for a handful of invitees, and for the rest, the closest they get to the president is looking at him on the stage. However, the event itself still carries massive import, not only for the meetings and conversations that happen in the White House at the time, but because an invitation to the White House acts as a lynchpin for the week. Kevin Sullivan, a first-generation Irish American who worked in both Clinton administrations and briefed the president ahead of his visits to Ireland, has remained committed to Irish affairs. He explains the value of the day now.

'It still has the same level of meaning. Everyone who is an Irish politician comes over for the American Ireland Fund dinner [usually on the evening before the White House events] so there's sidebar conversations before that day. Then there's sidebar conversations during the dinner. There's sidebar conversations the next day around the White House event. So Irish political leaders, North and south, make a point of connecting America with Ireland.'

He gives an example of how that manifested in March 2019 as the clock was ticking down to the original Brexit deadline of 29 March.

'Arlene Foster and some of the DUP came in and they wanted to talk to members of the Republican Party and the Trump administration, but Mary Lou McDonald of Sinn Féin came in and she wanted to talk to other people. So [former Republican congressman] Jim Walsh and I held a meeting with her, arranged about ten or twelve people to meet with her. I think the Irish find it useful still to come to Washington, to have those sidebar conversations, even if they're not necessarily taking place in the White House itself.'

With each passing US presidential administration, though, the events to mark St Patrick's Day at the White House are thrown into uncertainty. There is, of course, no obligation on any US president to host an Irish taoiseach, never mind spend practically the whole day with them.

Ireland's ambassador to the US from 2013 to 2017, Anne Anderson, explains that the celebration is important of itself to mark the national feast day, but also for how it is perceived for the rest of the year.

'You cannot overestimate the St Patrick's Day effect because it's not just what it delivers for us on the day, but the signal that it sends that we are a country with special status with the administration, that we are a country to be reckoned with. There is a spillover from St Patrick's Day that lasts year-round, and that is why that access is so precious. And of course other people will give you access because they know that your country has weight, and that leads them to be interested to spend time with you, to talk with you and hear you out.'

That is US politics, and in particular Washington DC politics, to a tee. It's all about who you know and how connected you are. This claim that Ireland has on the White House and Capitol Hill and all the power players for that one day a year generates returns all year long.

But it cannot be taken for granted. While the invitation is traditional, it is not guaranteed. Each time there is a change of administration, the White House elements are at risk of cancellation, and each time there is a change of power in the House of Representatives, the Speaker's Lunch is also at risk.

Given that then candidate Donald Trump threw out the political playbook on the campaign trail in 2016, and now President Trump continues to operate a freewheeling administration, there was no guarantee and, in particular, a good deal of nervousness as to what would happen in March 2017. Would President Trump, so keen to do things his way, keep the traditional celebrations and bilateral meeting?

At the time, there was an embassy-wide effort to secure the traditional events among the Irish diplomats in Washington DC, but ultimately the responsibility stopped with Ambassador Anderson. On what would be the last posting of a long and stellar career in the Department of Foreign Affairs, under no circumstances did she want to be remembered as the person who 'lost' the celebration. 'It gave me many sleepless nights; I was extremely concerned about it.'

Then taoiseach Enda Kenny was the first European leader to speak to president-elect Donald Trump by phone in the hours following his November 2016 election victory. That phone call came about because of much back-rooming by Irish diplomats on the night. An Irish American friend of one Irish diplomat happened to be working at the same law firm as one of Trump's closest consiglieri, former Mayor of New York, Rudy Giuliani, and the diplomat's friend convinced Giuliani to get Trump to take the call from Ireland. And so, in the early hours of 9 November 2016 Trump and Kenny spoke.

'You know, that phone call required a degree of agility because, in all honesty, like so many other people, we had thought a more likely outcome was President Hillary Clinton, so we had set up phone calls with her team,' explains Anderson. 'But in any event, that phone call took place and Enda Kenny said publicly that the president had invited him to the White House. The problem was it was a one-on-one call, so absolutely nobody in President Trump's entourage had any knowledge of this conversation or was willing to make any commitment in relation to St Patrick's Day. I was calling everybody, but I was getting no traction. There was no point in insisting this was all about long-established precedent because this president takes pride in not respecting precedent. He's not keen on socialising; he doesn't enjoy partying.'

So a one-on-one private statement from the new president, particularly this new president, to the then taoiseach, was not solid enough to protect what is one of Ireland's foreign policy cornerstones. But Anne Anderson, one of Ireland's most accomplished diplomats, was not going to rest until she had something more concrete.

'I distinctly remember we were within a few days of the inauguration – you know, that is very close to St Patrick's Day – and there was still no commitment from anybody. I was at one of these black-tie banquets for the president a few nights before the inauguration and I felt at this stage, having tried all other avenues, I had no option but to try to buttonhole the president-elect himself. So I waded through the crowd and introduced myself and, without any preliminaries, launched into how much we were looking forward to seeing him in the White House. I was fairly comfortable that we would have the meeting in the Oval Office – that was the courtesy I was fairly sure would be extended. But I wanted what sets us apart – the party. And so I told him how much we were looking forward to the party and the reception. I have to say, he looked slightly bemused, but he politely said he was looking forward to it as well.'

Another promise from the president – but were there any witnesses this time?

'Reince Priebus [who would become the new president's first chief of staff] was within a few feet of us, so I immediately put my hand on Reince Priebus, who looked equally perplexed, and told him, you know, the president was really insistent that this party would go ahead. So then he put me in touch with his chief of staff and everything happened from there.'

A few days later the invitation would be solidified beyond the point of no return at the White House press briefing. On then press secretary Sean Spicer's second day in office, this Irish journalist was called on for a question. I asked about the new president's intended tax policy towards American multinationals based in foreign countries, for example Ireland, as this was of great importance to Irish interests. And then I managed to get a second question in too. Unaware of the diplomatic efforts to have the invitation confirmed, I asked whether President Trump was planning to invite the taoiseach to the White House for all the traditional celebrations for St Patrick's Day.

Spicer laughed and said, 'I'm going to get on that right away. Thank you for bringing that up. That's on the list!' He added to the aides, 'Can you write that one down?'

The following day Spicer returned with the confirmation. 'It's an issue that's near and dear to me,' Spicer said. 'I was asked yesterday about the status of the invitation to Prime Minister Kenny of Ireland to visit the United States on St Patrick's Day and I'm pleased to announce that the president has extended that invitation. It happened during the transition period. We look forward to the prime minister attending.'

And so it was announced in public and the meeting was safe. And despite the concerns about the one-to-one nature of the Trump–Kenny conversation on the night of the election, the president seems to have viewed that as the point at which the invitation was issued. There are frequent criticisms that Trump can say things he doesn't mean, but on this occasion, it seems he considered the invitation as a 'done deal' from the outset. But as former ambassador Anderson recounts, the risk of cancellation, or minimisation, at the time was real.

'The reality is I did not want us to lose any element of the St Patrick's Day menu because any erosion of those elements would erode our visible claim to very special status, and the chances would be that we would never get it back again. Any interruption could break that precedent, and it really was at risk at the time of the transition. We were the first party that was held in the White House and they scarcely had catering staff in place at that stage.

But it isn't simply because we want a drink in the White House – it is because of what it signifies and what it means in terms of the year-round spill-over effect of St Patrick's Day.'

It wasn't a risk that existed just because of President Trump – it is one that is there every four years, regardless of who takes over.

'It's going to be more challenging, because a president who had no connection to Ireland could legitimately ask herself or himself on arrival in the White House, "Well, why this whole day devoted to Ireland? Yes to the Oval Office meeting and the exchange of shamrock, but nothing more." Particularly given the fact that there's only two days in the year where the president is guaranteed to be on Capitol Hill: the State of the Union Address and St Patrick's Day. And especially if you had a situation where you had a level of discomfort between a president and a Speaker, a president might well not want to be responsive to an invitation from the Speaker. And then going the extra mile to have that evening reception and what that signifies is really very important. But it will continue to be challenging to maintain into the future.'

The current ambassador, Dan Mulhall, says he's not sure if people in Ireland realise just how sacred, and challenging, the day is.

'I daresay there are people at home who think that we just waltz into the White House every year and they're delighted to see us. Contact with the White House and with the State Department starts before Thanksgiving and goes on until the time of the visit itself. A lot of legwork goes into getting the whole thing set up. It's true that there is a positive predisposition towards us and towards the St Patrick's Day celebrations, but we do have to continually work hard to ensure that the St Patrick's Day routine doesn't suffer.'

Sean Spicer, former White House press secretary and one-time communications director for the Republican National Committee, agrees. He's so proud of his Irish roots that long before he was in the White House, he would wear a green shamrock-emblazoned suit to do political television appearances. His great-grandfather, born in Liverpool to Irish parents, before moving to New York, was awarded the US Medal of Honor for sweeping for and disabling mines in the Spanish-American War.

'To be in a room like that with the taoiseach, the Speaker, and the leaders – it's not that big a room when you consider the amount of power and importance there – it's pretty awesome. The whole day is just pretty phenomenal.

'The thing that is truly unique in America is that there is not another country that gets the attention that Ireland does. So our entire country stops and celebrates St Patrick's Day in a way that we don't celebrate any other culture or country. The Greeks try it a little, but it's not by any stretch close to the celebration that occurs for St Patrick's Day and the recognition that Ireland has.'

The Greeks would probably disagree that they 'try it a little'. Since 1987, there has also been an annual White House reception held for Greek Independence Day. It falls on 25 March, but like the St Patrick's Day celebration, the White House moves the day of the official reception to suit its own calendar. The Greeks get a presidential proclamation, just as the Irish do, but the tone and focus is often a little different. As already illustrated, the Irish statements focus on culture, immigration and human connections, while the Greek–US bilateral relationship is spoken about with more geopolitical overtones. For example, the 2019 edict spoke of the relationship as affording 'many opportunities to support partnerships and initiatives that address the areas of defense and security, law enforcement and counterterrorism, and energy security and diversification'. The Greek prime minister is not usually in attendance, and in 2019, President Donald Trump was joined by the Archbishop of the Greek Orthodox Church in America, Archbishop Demetrios. (He has since resigned the position.)

The last time a Greek prime minister attended was in 2010. This has prompted Greek commentary that the Greek government should treat the event more like the Irish government does,[17] using the opportunity to build a relationship with the occupant of the White House. It has also been pointed out that Ireland does not send a cardinal, so the Greek American community should not send a religious leader, rather a political one.

So while the Irish celebration may be the envy of other diaspora communities, fears remain over its continuance. Spicer does not think it would ever make political sense to end the tradition. 'I think it has become ingrained in American culture and the presidency. I would have a hard time thinking that any president, not only would want to [cancel it], but would risk facing the political backlash from not doing it.'

It is an important event from an Irish foreign policy perspective, but is it just a fun photo opportunity for the Americans?

Speaking of the 2017 event, the first which involved President Trump, Sean Spicer says it was viewed as something of a jolly day. 'It's a fun tradition that has evolved. Obviously, there's preparation. Whenever two leaders meet, there's always gonna be preparation, but I think the nature of the relationship with Ireland is such that it's a very pleasant relationship. There's no huge bilateral issues to be resolved in these meetings, which is not always the case. With Ireland you're holding the meeting because you've got a great relationship, and you're hoping to make it stronger.'

17 See for example this opinion piece from Tom Ellis, editor-in-chief of the English language Greek news outlet *Ekathimerini* [http://www.ekathimerini.com/238957/opinion/ ekathimerini/comment/a-greek-prime-minister-at-the-white-house]. 'The idea is for the country's prime minister to participate in the event, along with representatives of the Greek American community, in the way that it is done by other diasporas – for example Irish Americans. This would offer an opportunity for informal contacts between the leaders of Greece and the US, which would not necessarily have the form of an official meeting with the participation of advisers, but would nevertheless offer an exceptional opportunity to build a personal relationship. Also, in this context, the Greek American community has to reach a point of maturity and be represented by a secular leader who will be voted or selected by prominent members of the community, and not by the archbishop. The Jewish and Irish communities are not represented by a rabbi or cardinal.'

Although the Americans may have viewed that 2017 meeting as one that did not have 'huge bilateral issues to be resolved', the Irish side still had plenty of points they needed to raise with the new president and his team. While the Irish diplomats and then taoiseach Enda Kenny had experience in going to the White House for this day, it was the first time for the new Trump administration. This threw up more difficulties to be navigated – how would this new playbook-shredding president approach the meeting? How would Enda Kenny balance preserving the US–Ireland relationship while still reflecting the views of many Irish people on how the president had behaved during his first few months in office?

Then ambassador Anne Anderson describes the visit every year as 'a balancing act'. But she says it's 'sometimes a more delicate one than others for a visiting prime minister'. She explains: 'It's obvious that was a challenge during the first year for the Trump administration because any Irish taoiseach has got to be true to the citizens, faithfully and accurately representing the interests and the values of Irish people. On the other hand, you are interacting with a democratically elected president of the most powerful country in the world, and there is the dignity in the office that needs to be factored into that exchange. There's the fact that you're a guest in the White House. So finding the appropriate language to respect that balance can be quite difficult, and it's a question of contextualising what you say, finding the right nuance in the presentation.'

Sean Spicer remembers a cosy atmosphere. 'It was a very warm, very pleasant meeting. With an Irish–US bilateral, there's no acrimony. It's pleasant, and it's welcoming, and it's fun. You know there's an event coming up afterwards, a celebration, which makes everything surrounding the meeting enjoyable. It's not negative. It's fun and exciting, and everybody's enjoying themselves. And, especially as America has grown more partisan, more ideological, it's an opportunity to have one of those rare moments of celebrating culture across party lines.'

The Irish side took a strategic approach that was different from that of previous years. Anne Anderson explains: 'Right up until the meeting itself we were discussing nuance, tone, and I was not signalling in advance on that occasion. Normally, of course, meetings between the president and the taoiseach would be very carefully prepared in advance with a lot of questions already discussed. Usually you just fix an agenda, and you give a broad sense of what you expect principals to say on each side because that's how good substantive meetings are best prepared. And if you want an outcome, you will have covered the territory well beforehand. So in advance of each of the meetings with President Obama there would have been good signalling going on between myself and counterparts in the US administration. I think on both sides, for the first meeting with President Trump, we judged it better to leave it to the principals to set the tone themselves on the day.'

She describes President Trump's approach as 'freewheeling' and 'organic', that he spoke a lot but also listened a lot. President Obama by comparison gave the impression that he had 'mastered every aspect of the brief'. Those meetings, says Anderson, were 'friendly, but very cerebral, very structured, very disciplined'.

Anne Anderson's counterpart at the time, US Ambassador to Ireland Kevin O'Malley, was also present for those meetings.

'I would say that of the two meetings I participated in with Enda Kenny and Barack Obama, they were very pleasant but serious. Each person, the president and the taoiseach, literally pulled out a list of things they wanted to discuss, and they discussed them in detail. The taoiseach, for example, brought up the question of Irish immigration into the US and the undocumented Irish. In my two meetings, that was the first thing that Enda Kenny brought up. President Obama brought up other topics. We talked about taxation and corporate inversions and things. So it was a totally substantive meeting, followed by a press conference in which both leaders summarised what had occurred and answered some questions.'

'Press conference' is a bit of a generous term for what happens next. It's standard when the US president is hosting another world leader that members of the media are invited into the Oval Office to take video and photos and try to ask questions, and the day of the Irish visit is no different. The White House press pool, including the permanent Washington correspondents for RTÉ and the *Irish Times*, and the visiting Irish media, are all corralled outside the Oval Office waiting for the bilateral meeting to end. As soon as the leaders are ready for the media, a White House communications aide opens the door and the photographers, camera people and journalists rush in. There is about 30 seconds to get a spot and get positioned before the two leaders start speaking. As soon as they are finished, there is about 30 seconds during which the media members are all whipped back out again. There is no time allocated for questions, but reporters shout them nonetheless; sometimes they are answered, sometimes they are not.

Former US ambassador Kevin O'Malley remembers his first St Patrick's Day Oval Office experience clearly. 'Vice President Biden was there. This was my first meeting, and when the meeting ended, President Obama asked the taoiseach, "Is it okay to bring in the press?" And the taoiseach said, "Sure." And Vice President Biden put his arm on my shoulder and said, "Ambassador, I think you and I should move." So we got up, and I wasn't sure what he meant by that, and we moved to the back of the room behind the president's desk, and then they opened the door, and all you guys ran, R-A-N, ran in and set up your stuff, and I told the vice president later, I said, "You saved my life. These guys would've killed me." The fight to get into the front row! I don't know what the deal was, but I actually kind of laughed. To me, being in the Oval Office was more a genuflection event, and you guys were making it like a sprint. So it was really kind of funny.'

But apart from moments of levity, the former ambassador says that the Obama administration took the meetings very seriously. He can't speak for what happens currently, as he is no longer involved with the US State Department, but he says that during his time this was the case.

'By the same token, I wouldn't want to suggest that whatever kind of preparation that went in for President Obama to meet with President Xi of China was the same. It's very pleasant work, for the people who have an opportunity to serve at the US embassy in Dublin, and the people who work on Irish affairs at the State Department. It's taken seriously, but in a very pleasant way. We take issues in Yemen very seriously, but there's nothing pleasant about it.'

As noted, striking a balance between being serious and pleasant posed a challenge for the Irish team for that first meeting with President Trump. After all the effort to secure the meeting, Ambassador Anderson did not want it to backfire with an approach that might have offended the notoriously thin-skinned American president.

'I think Enda Kenny did very well on the day because, particularly at the Speaker's Lunch, he managed to convey a very clear message on immigration. But the fact that he had established a kind of rapport with President Trump – Enda Kenny was a very friendly, tactile person – gave him a bit of leeway to convey a message that was indeed very clear and created a few headlines in the States and around the world, without the president taking offence at what had been said. As we have all witnessed, this is a very thin-skinned president, so he had to factor that into the calculations.'

She's referring to remarks that the then taoiseach made about the plight of the undocumented Irish migrants in the United States. At the Speaker's Lunch on Capitol Hill, Mr Kenny said: 'It is not for me to insert myself into your political debate but I would highlight the plight of our undocumented community of up to 50 thousand hardworking Irish men and women, who are tax-paying contributors to this, their adopted home. Many of them came here decades ago, in a different era, and have been caught in limbo without any path to regularise their status. On this day when we remember St Patrick, himself an immigrant twice over to our shores, I urge you to look sympathetically at this issue.'

Later, at the presentation of the traditional bowl of shamrock, he followed up those words.

It's fitting that we gather here each year to celebrate St Patrick and his legacy. He, too, of course, was an immigrant. And though he is, of course, the patron saint of Ireland, for many people around the globe, he is also a symbol of, indeed, the patron of immigrants.

Here in America, your great country, 35 million people claim Irish heritage, and the Irish have contributed to the economic, social, political and cultural life of this great country over the last 200 years. Ireland came to America because, deprived of liberty, deprived of opportunity, of safety, of even food itself, the Irish believed, four decades before Lady Liberty lifted her lamp, we were the 'wretched refuse on the teeming shore'. We believed in the shelter of America, and the compassion of America, and the opportunity of America. We came, and we became Americans.

We lived the words of John F. Kennedy long before we heard them: we asked not what America could do for us, but what we could do for America. And we still do. We want to give, and not to take.

—

Anne Anderson's successor, Ambassador Dan Mulhall, feels both the US and Irish sides have now settled into the meeting. His first St Patrick's Day as Ireland's ambassador to the US coincided with

Leo Varadkar's first as taoiseach so, again, there was a new dynamic to be tested.

The freewheeling approach required in the first year with President Trump has been replaced with an agenda in the two years that Dan Mulhall has been there with Taoiseach Varadkar.

'It's as serious as any meeting I've been at in my forty-year career,' says Ambassador Mulhall. 'I mean, obviously meetings, generally speaking, take an hour. The one in the White House takes more or less an hour. We go through the agenda. Broadly speaking, everything on the agenda is dealt with, and other people in the room get involved as well, chip in. But the discussion seemed to me to be fairly normal in terms of the kind of high-level discussions I've attended in other parts of the world.'

Speaking with the president's chief of staff, Mick Mulvaney, a couple of days before the 2019 meeting dispels any notion of a lack of serious intent. 'Oh, the boss knows the taoiseach is coming this week.' He adds that President Trump has just been on to him about immigration reform, to try to have something positive to report to the taoiseach when they meet in the Oval Office. 'It's not just a photo op. Absolutely not.'

Mulvaney says it's a big deal for him, having an Irish background, to take part in the ceremonial events around the day, but the bilateral Oval Office meeting is always an important one. Mulvaney was the director of the Office of Management and Budget but took over as chief of staff when his predecessor, John Kelly, left. Mulvaney says they were the strongest proponents of Ireland around. 'John Kelly and I were sort of the head lobbyists for Irish issues in the United States, at least in the administration. We worked very closely on immigration. So now it's just me but we'll continue to push the issues.'

It can be rare for President Trump to mention Ireland publicly and not mention that he has a property in Doonbeg, County Clare, and the St Patrick's Day Oval Office meeting seems no exception.

But, as Ambassador Mulhall explains, 'The taoiseach doesn't play golf, so that doesn't really arise in conversation with him. But I've sat together with the president now a couple of times and he does mention his affection for the property he has in Doonbeg. I think he has a genuine enthusiasm for it.'

On his way from his previous position in London, before arriving in Washington DC to take up the post of Ireland's ambassador in August 2017, Dan Mulhall made a point of stopping off at Trump International Doonbeg to play at the golf course. And in 2019 a promise from the president to travel to his resort in Doonbeg was realised a few months later with a trip to Ireland in early June. Ambassador Mulhall (and plenty of others) played the course again, this time alongside the president.

The visit had caused diplomatic headaches in advance as the president wanted to be based in Doonbeg, but protocol would not permit either the taoiseach or Irish president to hold a formal meeting at a commercial property. A compromise was found, and Taoiseach Leo Varadkar met President Trump for a bilateral meeting at Shannon Airport. There were some sizeable protests, but the president was kept far from them, flying in and out of Shannon to attend D-Day commemorative events in France.

Little of substance came from this bilateral meeting, but nonetheless it was important from an Irish soft power perspective. The American president promises he will come to visit – and he does. There may not have been interesting images of the Irish parliament or geopolitical negotiations beamed around the world, but there was plenty of green scenery and Irish hospitality on show as the president's sons, Eric and Don Jr, pulled pints and enjoyed banter with the local people. And these are the images that a great deal of Americans – whether of Irish heritage or not – associate with Ireland.

—

Obama's chief speechwriter, Cody Keenan, was in charge of writing no less than 24 sets of Irish remarks over his time with the president, three every St Patrick's Day – one set to the media at the Oval Office bilateral, another at the Speaker's Lunch on Capitol Hill, plus a third set at the evening reception in the White House. 'There is no lobby in Washington that commands three sets of remarks from the president once a year. It's wild.'

Keenan has Irish heritage 'way back'. His Irish ancestors came over in the 1700s – he doesn't know exactly when or from where, but he doesn't particularly care. There was a Patrick Keenan in his family who left Cork seven generations ago. On his mother's side, there was also an ancestor who came to the US from Dublin seven generations ago. He says he didn't grow up with his family instilling any sense of being Irish, but they would go to their local St Patrick's Day festival and their local Catholic church. It was only when he went to work on Capitol Hill for the late, great Friend of Ireland Senator Ted Kennedy that he became fully engaged in Irish affairs and more interested in his own heritage.

Fast forward six years and he was working as a speechwriter in the Obama White House, and somebody had to write the remarks for St Patrick's Day.

'We'd been in the White House for less than two months, and I basically begged Jon Favreau [then chief speechwriter, a job Keenan would eventually take over] "can I write the St Patrick's Day remarks?" He's like, "Yeah, sure." I don't think he took it super seriously because he wasn't Irish at all. So you do have people that just look at it as a fun day – it's St Patrick's Day, wear your green – but I was in the office for three nights straight, reading books, I consumed entire books of W. B. Yeats's and Patrick Kavanagh's poetry, I was doing so much research, trying to make these amazing speeches ... I remember, at one point, Jon was, like, "Dude, these are way overwritten." But I remember Denis McDonough [chief of staff] and Samantha Power [presidential adviser and US

ambassador to the UN] were, like, "These are amazing." So we had
Denis and Samantha and Tom Donilon [national security adviser]
and Vice President Joe Biden and me, and I'm sure I'm forgetting
a ton of other Irish, and we took it super seriously. As did the
National Security Council. Ireland is an ally and they approached
every kind of ally with a sense of seriousness.'

The case of Samantha Power in the context of Irish American
politics is worth noting. She was born in Dublin and moved to the
US aged nine, and was a journalist and academic before serving
in Barack Obama's cabinet as US ambassador to the UN. In that
role, she was unapologetic in her need to be American and put
US interests above all else. However, she did say that she used
the experience of coming from a small country like Ireland, and
being an immigrant, to inform her viewpoints and decisions at the
United Nations.

Having his own Irish connection meant Cody Keenan took the
speeches he wrote more seriously than if they were remarks to be
drafted for the visit of any other foreign leader.

'If it was Greece or Argentina or whoever, I didn't care as much,
I could let another speechwriter handle it, but the Irish, for me, is
our annual state visit from our friends. It was a blast, you know? It
was always a fun day. We dyed the White House fountain green –
we were the first to do that. But we had that Chicago connection,
too, where the river is dyed green. It was a blast; but I always took
it very seriously. So did the president, to his credit. He took every
speech seriously but these he cared about because he knew his
Chicago constituency, knew the people in the White House.'

Life in the White House operates at a frenetic pace and any
president going in to meet any taoiseach may have just been dealing
with some sort of geopolitical disaster in the Situation Room or an
economic crisis or any number of things. Cody Keenan says there
would be a team of people to prepare Barack Obama ahead of the
St Patrick's Day celebrations when he was president.

'It's such a big operation and everyone's so good at their individual roles that the president goes about his day, and when he shows up, the remarks are there, the bowl of shamrocks are there, so he didn't have to worry about it too much. I do remember that first year I was so excited for the whole thing – again, we'd been there less than two months. I went up into the Roosevelt Room just to watch the shamrock ceremony with – I think it was Brian Cowen that year. It was always great fun. Then you go over to the Capitol for the Speaker's Lunch. The reception at night was always rowdy – it was one of the rowdier parties every year. I think President Obama enjoyed that – he got a good kick out of it. He's not the type to stay and party with people. He just kind of does the remarks and gets out. He'd always say, "Stay, have fun, don't break anything."'

The White House evening reception can indeed be a pretty rowdy affair. What you don't see watching on television is the sometimes significant number of people who can't fit into the East Room for the speeches and are spilled off in a corridor to the side, or who choose to stay in front of the buffet (free-flowing champagne, whiskey, stout, beef canapés and green cookies and desserts) enjoying the hospitality. A select souvenir is, of course, the napkins, causing President Obama to joke on one occasion about having replaced the cloth napkins with paper ones because so many were disappearing as keepsakes. On another occasion, the jovial chatter from those not in the room was so loud that Congressman Joe Crowley, a tall man who could see over most people's heads, turned around and issued a loud *shhh* in the hope of shaming the partygoers into listening more intently to the president.

Whatever might be discussed at the bilateral meeting behind closed doors between the president and taoiseach of the day, in the Obama administration there would always be official talking points from the National Security Council – in charge of US foreign policy – that needed to be inserted in the speeches.

The remarks, including those given by President Obama on his visit to Ireland, were always about an Irish audience and a domestic US audience, says Keenan. In the modern age of social media and international internet broadcasts, one can't assume remarks will stay in the room. 'When you're writing, you never write for one specific audience. You know that everybody's listening.'

Keenan says Barack Obama also took his Irish heritage seriously. Often the "O'Bamas of Moneygall" is used as a joke and people snigger about busloads of American tourists stopping at Barack Obama Plaza services on the M7, but Cody Keenan, who still works with the former president, says Barack Obama took his Irish ancestry to heart. He was genuinely moved by the trip to Moneygall.

'He told me, "I went in that house, I walked around on the same floorboards that my great-great-great-grandfather walked on, and that's a pretty powerful thing." One thing he's never told me, I'm just divining on my own, and I could be wrong but he only met his dad twice. In America, we tend to look at anybody with black skin of any shade as black. But he grew up with his white mother, and his white grandparents. He's obviously always grappled with questions of race, and identity, and wrote a book about it, but when people pooh-pooh the Moneygall connection, I mean, that's the side of the family he grew up with. Here in the office, he's got this drawing somebody gave him of Falmouth Kearney up on the office wall.' Kearney was Obama's great-great-great-grandfather who genealogists have traced leaving Moneygall in County Offaly for New York in 1850.

President Obama spoke of his feelings at being in the cottage when he addressed the Irish people from a stage in College Green in 2011. 'Standing there, in Moneygall, I couldn't help but think how heartbreaking it must have been for Falmouth Kearney and so many others to part; to watch Donegal coasts and Dingle cliffs recede; to leave behind all they knew in hopes that something better lay over the horizon.'

That connection President Obama feels is what his ambassadorial appointee Kevin O'Malley describes as a connection 'that is not a political tie. It's a tie at a cellular level.'

It's clear he relished his time as US ambassador to Ireland, but he enjoyed some parts of the role in particular. O'Malley was not a career diplomat but was an organiser for Obama in Missouri – he's quick to point out that he was not a huge donor himself, but was good at getting donations from others.

'One of the thrills of being ambassador is being able to invite people to the White House. I insisted with President Obama – well, I shouldn't say I insisted: I asked the president for a larger allocation of tickets than previous ambassadors had enjoyed and tried to spread that out among people that I knew were contributing to the Irish American connection. So that was fun. To be an American and be invited to come into the Oval Office, let alone to have a substantive meeting with my president and the taoiseach, was remarkable and an experience I don't think I'll ever forget.'

While the hottest civilian ticket in town that day is to the White House evening celebration, the hottest political ticket is to the Speaker's Lunch on Capitol Hill. That's a more recent tradition that began while Ronald Reagan was president. For this lunch, tables are squeezed into what is usually a hallway near the Speaker's office that is transformed into a lunch venue, but with a maximum capacity of 110, which is never enough. It is truly bipartisan and congresspeople and senators from both parties vie for tickets. The president, vice president, Speaker, taoiseach and respective Irish and US ambassadors are guaranteed seats, but that's about it. After that, usually the chairperson of the Friends of Ireland caucus will try to accommodate those who are active within the grouping, but anyone with an Irish last name, whether they've nurtured the connection or not, tries to get into the room.

Current chairman of the Friends of Ireland caucus is Massachusetts Democratic Congressman Richie Neal. 'The caucus members definitely get preference, but it is a hot ticket.'

Given his own immigrant history, he never ceases to marvel at the day. 'It's a reminder of my own steep climb – to have immigrant grandparents and be sitting in the room with the president and the Speaker of the House, the prime minister of Ireland. Not bad. How many other countries the size of Ireland with six million people North and south command the presence for an entire week in Washington? Audience with the president, lunch with the Speaker of the House. All the political parties will be represented throughout the week.'

The festivities are not just about wearing green and being seen at these high-level events. Republican New York Congressman Pete King says, over the years, the meetings played an important role, particularly during the Northern Ireland peace process.

Democratic Congressman Brendan Boyle, from Philadelphia, says it is important that domestic politics and feelings about individuals don't get in the way of the relationship and the tradition.

'It's been wonderful the way the Speaker's luncheon, and the White House passing of the shamrocks has continued uninterrupted year after year – really amazing. I think folks are conscious that if there was ever an interruption in that and it was lost, it would probably be lost for good. It's a realistic fear, because the number of similarly sized countries to Ireland that would give their right arm to have that level of influence and access is something that makes me pretty proud to be Irish American.'

He thinks the fact that President Trump kept the full day's menu of festivities to mark St Patrick's Day is a good sign for the future. 'I really give him credit. I know that given how unpopular Trump is in Ireland, in pretty much every country on earth, it was not necessarily the most popular decision for Enda Kenny, and then for Leo Varadkar, to meet with Trump and to attend that ceremony. But I think it's incredibly important that they did, and they made that decision, I believe, based on the national interest of Ireland and not necessarily their own political interest.'

His party colleague, Massachusetts Democratic Congressman Joe Kennedy III, agrees. 'It highlights the ties between our countries. It highlights the commitment that our countries have made to each other, regardless of party, regardless of who's in office, who is in power back in Ireland or here at home. It is something to be nurtured and respected and it would be a shame for anybody to do any damage to that.'

—

As chief of staff to the president, Mick Mulvaney has one of the bigger and brighter offices in the West Wing. He's at the other end of the corridor to the Oval Office, next to the vice president's office and around the corner from the National Security Council adviser. His office comprises a 'standing-up' desk in the corner, a large boardroom table and a cluster of sofas and armchairs around a fireplace. Above the fireplace there's a painting of President Abraham Lincoln. In a little cubby behind the door there's a kettle, some cups, coffee, creamer and boxes of Bewley's tea. Mulvaney points out that he bought these teabags himself on a recent trip to Inis Mór. The carpets and upholstery are beige-toned and there are copies of the *Financial Times* and the *Wall Street Journal* on the coffee table.

Mulvaney was a four-term Republican congressman who joined Team Trump after the election. He was the director of the Office of Management and Budget (OMB) following President Trump's inauguration and is now the president's chief of staff.

We're talking about the painting of President Lincoln and I joke that the first time I interviewed him after he took a role in the

Trump administration, his aides were hanging his own personal paintings and maps in his office in the Eisenhower Executive Office Building (EEOB) where his OMB office was. The West Wing office is already furnished, but he says he wants to take down the Lincoln painting as soon as he can. He comes from the border of the North and South Carolinas, south of the Mason–Dixon line, where Abe Lincoln is not a popular man because of how the area was pillaged during the American Civil War.

It's a few days before the traditional White House St Patrick's Day celebrations so there is much talk of Ireland in the air. His aides are recommending he catch up on shows like *Derry Girls* and *Peaky Blinders*.

Mick Mulvaney considers himself Irish American, is a friend of the Irish embassy and, while a congressman, was a supporter of the Friends of Ireland caucus on the Hill, but he says his exact Irish heritage is 'not definitively determinable', the reason being that it was quite a few generations ago and there was a problem with the paperwork.

'My great-great-grandfather forgot to provide his middle name or initial when he emigrated. He emigrated through Canada. Six Matthew Mulvaneys that emigrated in the year that we know he left went through Canada. Five of them are from Mayo; one of them is from someplace else. So we assume, statistically, that my family is from Mayo, but we don't know for sure.'

As a result, he doesn't have relationships with any family members who may be still in Ireland, but the pride in his roots is there, nonetheless. 'Mostly we just tell everybody I'm from Mayo and that's the end of the conversation.'

Despite the slightly fuzzy connections, he was raised in Charlotte, North Carolina, with an Irish upbringing. His family were members of the Irish society there, and they went dancing and, later, drinking in the local Irish social club. His niece did competitive Irish dancing for fifteen years.

Unlike others, he doesn't think having Irish blood had anything to do with his move into life in public service. However, he is a staunch Catholic, and he feels being an Irish Catholic was more of an issue. He says he was the first Catholic member of Congress ever to be elected in South Carolina. During his campaigns for the US Congress, and before that the South Carolina state legislature, he never stressed that he was Irish or Catholic. He told people he was Catholic 'to get it out of the way' – especially as, he says, there are 'some old prejudices against Catholics in the South. They are in large part fading away. But less than 1 per cent of the people are Catholic in the state I grew up in – North Carolina. I live right on the boundary, so I go back and forth between North and South Carolina.'

But, he says, the number of Catholics in North Carolina has grown dramatically over the course of his lifetime. When he was going to school there were only two Catholic high schools in the whole state of North Carolina. Now there are 48 serving nearly 12,000 students. There is also a big Irish community in North Carolina now, and one of the country's biggest parades runs in Charlotte each year.

Nevertheless, as a veteran of elections, Mick Mulvaney does not consider there to be a sort of Irish voting bloc in the Carolinas, or even one to speak of nationally. He believes there are Irish voting communities in some of the traditional cities like Boston, New York, Chicago and even possibly Minneapolis and Milwaukee, Detroit and Cleveland. But that doesn't exist once you leave the larger cities of the north-east and Midwestern states. He thinks it has more to do with where you are from than your ethnic Irish background when it comes to politics.

'If you're Irish and you're from New York you're a Democrat. If you're Irish and you're from Texas you're Republican'.

Religious preferences can change that, though, says Mulvaney. 'Keep in mind that not all Irish are Catholic, but if you want to paint with a broad brush, Catholics would be pro-life. Generally speaking, they might come down to the Republican side of the ledger on that issue. But there's other ones who might not go that way.'

He thinks economics matters too, and there are probably more Irish American Democrats than Republicans, although there are no statistics on the matter.

'It's a sort of working man's, working woman's, blue-collar attitude. Even if the Irish have gone off and made a lot of money, they remember that they didn't always have it. Most of us came over poor. Even if you're just a couple of generations removed from a famine, maybe you'll sort of see yourself as more middle class, or upper middle class and not really elite. Maybe you're more inclined to like somebody like Donald Trump who did very well in the Midwest with working classes, did well with certain Democrats, just as Reagan did.'

He does, however, think there is 'a cultural thing that unites both the parties'.

'There's me and Pete King [Republican congressman] and then Brendan Boyle and Joe Kennedy [Democratic congressmen]. So there is that cultural sort of connection and oftentimes we work on Irish issues together across the aisle because of the cultural connection. Is that a voting bloc? No. Does that mean the Irish punch way above their weight? Yes, it absolutely does.'

All of a sudden, the secure hotline phone starts ringing, and I'm asked to step out of his office for a few moments. He's been waiting for a call from Secretary of State Mike Pompeo and this is it. At the time, the crisis in Venezuela was peaking and the US was pulling its last remaining diplomats out of the country.

Waiting in his antechamber for the discussions on national security matters to end, I notice that Vice President Mike Pence has his office next door. Pence has the most tangible connections to Ireland of any president or vice president in recent times. He speaks often of his Irish American roots and frequently tells the story of the role played in his life by his grandfather Richard Michael Cawley, who left Tubbercurry in County Sligo in 1923 for America, worked as a bus driver in Chicago and did not see his mother for 25 years. At his first St Patrick's Day White House celebration as vice president, in 2017, he spoke of his impressions of the Irish in America based on his own personal and emotional family connection. He spent many of his summers as a young man working in the west of Ireland and visiting cousins. He regularly cites his Irish heritage as the reason for his public service contributions.

'The truth is my grandfather was very typical of the millions that would come to these shores,' said the vice president. 'He embodied all that's best about the Irish – sturdy work ethic, faith in God, love of family, patriotism. And those are the enduring contributions of people of Irish descent in the history of this country. It's extraordinary to think of the contributions that the Irish have made. In every single American conflict since our Revolutionary War, the Irish people have enriched America in incalculable ways, and they always will.'

Pence says that he had the story of his grandfather in his mind when he swore the oath of office in January 2017.

'I had many things in my heart and mind that day. But my mind kept going back to my grandfather, to his courage to be able to leave everything behind, everything he knew – family, and hearth, and home – and come here because it was a future, because he believed in the American Dream. I was up on that platform thinking that Richard Michael Cawley and his courage is the only reason that Michael Richard Pence became the 48th Vice President of the United States.'

Vice President Pence says there is a legend in his family about how his grandfather came to leave Ireland for America.

'His mother walked him across the street from their modest home in Ireland, and walked him up the hill, and pointed in the direction of the Ox Mountains and to the west, and said, "You have to go to America because there's a future there for you.'"

However, Pence's own somewhat misty-eyed view of his family's immigration is often contrasted with the approach the administration he is a part of takes towards current immigrants. Coming to America to find your fortune and make a better life for your family no longer guarantees a welcoming reception.

A few minutes later, Mulvaney's call with Pompeo is complete and we're back talking about Ireland.

He does feel the Friends of Ireland caucus on Capitol Hill is an effective working group. And although he's no longer a congressman he is still working with them, a recent example being the efforts to get access to the E3 visa programme for Irish nationals.

The push for a special case for legal Irish immigration is based on the unintended consequences of reforms which date back to the 1960s. It was felt then that immigration to the US was too eurocentric, and an effort was made to rebalance the countries of origin of immigrants with the introduction of the Immigration and Naturalization Act of 1965. Until then, immigration was based on a national-origins quota system, whereby a nationality was awarded a certain number of visas based on its representation in the US census figures – clearly Ireland did very well. However, the 1965 Act ended that practice, and introduced a new one based on attracting certain skill sets to the US and reuniting families.

Senator Ted Kennedy was one of those who worked to end the nationality quotas, as did his brother President Kennedy, before his assassination. The consequence of this legislation, however, was a dramatically reduced number of visas issued to Irish citizens, a pattern that has continued for the intervening 50 years. Latest figures available for the issuance of Green Cards (2017) show that just under 1,500 were issued to Irish nationals in 2017, out of a total of almost 1.3 million – a mere fraction. Irish immigration campaigners point to a mismatch here. One in ten Americans claims to be Irish, yet that shared story cannot continue apace when modern emigration from Ireland to the US has stalled to a trickle.

The need to rebalance this is the impetus behind most of the lobbying for increased legal pathways for Irish people to come to the US. It's framed as the correcting of an unforeseen mistake, and is put forward as the reason why Irish immigrants could legitimately be treated differently from immigrants from other nations who do not have such a shared history. The issue of resolving those who are in the US illegally, without the correct documents, is more complex, and there is not a moral argument for dealing with the Irish separately from every other nation. A life 'in the shadows' is difficult, no matter where you are from. It cannot be argued that Irish people deserve special treatment.

With President Trump's well-documented views on comprehensive immigration reform, and a polarised Congress, any widespread policy changes are highly unlikely. The approach decided on by the Irish government and other campaigners was that the best immigration deal on offer at the moment is the effort to hive off unused E3 visas set aside for Australian nationals and allow Irish nationals to apply for those. The process began in early 2015, but was hampered by changing personnel and the short US election cycle. Ultimately a bill was tabled in 2018 and, although it was a bipartisan push, the effort failed.

The bill was co-sponsored by Democratic Congressman Richie Neal and by Republican Congressman from Wisconsin Jim Sensenbrenner. It was House Resolution 7164, 'To add Ireland to the E3 non-immigrant visa programme', and, as Neal put it, it 'sailed' through the house. It was passed by a voice vote so there is no exact breakdown of yeas and nays, but given that very little has passed through the US House of Representatives after the 2018 midterms, that was an achievement in itself. Because it was coming through Congress in the dying days of the last session, during the recess period immediately before Christmas, it required unanimous consent in the Senate to make its way on to the next stage – the president's desk for signing. That meant it needed all 100 senators. It failed when one senator – Republican Tom Cotton from Arkansas, a vehement opponent of immigration and a Trump hardliner – refused to vote for it. The president says he would have signed it, but that is easy to say after the fact. The clock ran out on the 115th Congress and there was no time to recast it.

However, all parties wish to have another attempt at getting the deal through.

Richie Neal drew up and retabled the bill on 30 April 2019: House Resolution 2418. He's confident it will get through the House – and with the current Democratic majority it should do so easily enough. Senator Tom Cotton remains a problem; however, as President Trump's chief of staff Mick Mulvaney says, the president himself is working on this and they're confident.

'It passed in the House on what's called suspension of the rules, which means you can do it without going through the committees, without going through debate. It's essentially for things that are so popular or so non-controversial that it would just move them through so we can move on to other things. And the Irish E3 visa in the House, at least at the end of last year, passed on that basis. So you see, it's sort of the fruits of that bipartisan support.'

Much has been made of this plan, but the scheme has been talked up, well beyond its importance. There are 10,500 E3 visas allocated to Australian nationals each year. This deal was negotiated alongside the Australia–United States Free Trade Agreement (AUSFTA) that came into effect on 1 January 2015. It's also understood by some to have been something of a quid pro quo based on Australia's participation in the Coalition of the Willing, backing the US in the Iraq War in 2003. Approximately half of that 10,500 allocation goes unused by Australians every year. It is not clear why Australians use so few of these visas, but it may be to do with the type of visas the E3s are. Initial pushback from the Australians has been smoothed over with the assurance that the Irish will only have access to the unused visas in the year after the Australians have indicated they don't need them.

An E3 visa will not do anything for undocumented or illegal immigrants already in the US, but neither will it do anything for someone wishing to come to the US to 'find their fortune'. The application criteria are quite strict: it is only for those who are professionals either by qualification or with lengthy experience in their position, considered to be about ten years. But crucially, an applicant must have a job offer from a US employer to begin the process – many US employers will not make a job offer to anyone who does not already have a work permit, so it's a chicken-and-egg scenario. The visa can be granted for up to two years and can be renewed.

These are not visas for those who might wish to do the traditional jobs of earlier Irish immigrants in the bars and restaurants and building sites of New York City, Boston and Chicago. These are for immigrants who have a certain standard of third-level education or qualifications in certain industries, all prescribed by the US State Department. So are there likely to be 5,000 Irish people who fit those criteria? Conservative estimates say it could benefit a hundred to two hundred Irish people a year – which, granted, is a hundred to two hundred more than now, but nowhere near the numbers that may wish to go. The E3 is what the US Department of Homeland Security defines as a non-immigrant visa. Statistics for 2018 show Australian nationals were granted 45,778 non-immigrant visas in 2018, including 5,394 E3 visas and 4,892 student J1 visas. Irish nationals were granted 16,047 non-immigrant US visas in 2018, 7,319 of which were J1 visas.[18]

18 US Department of State, 'Report of the Visa Office 2018', Travel.State.Gov. <https://travel.state.gov/content/travel/en/legal/visa-law0/visa-statistics/annual-reports/report-of-the-visa-office-2018.html>

In exchange for access to the unused Australian E3 visas, the Irish government agreed to a reciprocal arrangement that would ease restrictions on American citizens who wish to work in or retire to Ireland.

There was, perhaps, an understandable over-excitement about these E3 visas in the Irish media, the Irish American media and publicly in certain Irish government and diplomatic circles. Privately, most would concede these are more about delivering a 'win' than delivering opportunities for Irish people, given that applicants must have a professional qualification and a suitable job offer from a US employer in advance of applying.

If the retabled resolution makes it through both Houses of Congress, the president's chief of staff says it would arrive at an open door in the White House. 'The president is very interested in seeing it passed.'

At the other end of the hall from Mulvaney's office, President Trump was in the Oval Office on the phone to a Republican senator. In his early weeks and months (and possibly even longer) the president was charged with trying to ride roughshod over Congress, not appreciating that nearly half of them were members of his own party (even if he was a reasonably recent convert to the Grand Old Party, most congresspeople were lifers).

But lessons have been learned through failed efforts at overturning Obamacare and a lengthy government shutdown. In addition, the arrival of former four-term South Carolina Congressman Mick Mulvaney has clearly made a difference. Mulvaney knows how negotiating and vote-gathering on Capitol Hill work.

On this occasion, President Trump was on the phone to one particularly strong-willed senator, who liked to think he was going places and that preserving the backbone of true conservatism was down to him. Step forward Tom Cotton of Arkansas. The one US senator who blocked the E3 bill before Christmas 2018.

On the phone, on this particular Tuesday afternoon, President Trump was trying to persuade Cotton to support the efforts to extend E3 visas to Irish people.

That effort doesn't seem to tally with President Trump's view of immigration reform, but his chief of staff Mick Mulvaney insists that's the case.

'I'm not making this up. He was working on it, lobbying on it, just before you got here.'

If the president signs that bill, it would be one of the first positive immigration-reform acts that Trump would do.

Mulvaney agrees. 'Yeah, it's a big deal for him alright.'

However, he goes on to argue that, despite popular perception, President Donald Trump is in favour of expanding immigration possibilities into the United States. I point out that with his rhetoric and actions on the border wall, and on Latino and Muslim immigrants in particular, he does not come across as particularly pro-immigration.

'I think that's misunderstood internationally; it's misunderstood domestically too. But the president is not anti-immigration. In fact, he and I had this conversation just the other day. We're already running out of people, okay? We're nearly at, if not already at, full employment. Everybody who wants a job in this country has one. We have more job openings than we have people looking for work for the first time, I think, in recent memory. We may have had it 150 years ago, but we haven't had it in my lifetime. So we need more people, and he knows that. We're having a meeting this week, for example, on H2B visas. He has said many, many times that one of the reasons we need so badly to solve the illegal immigration problem on our southern Mexican border is because we also need to fix legal immigration. We can't fix legal until we fix illegal because people lose faith in the system.

'Ronald Reagan did that. Ronald Reagan famously swapped amnesty for border security. He gave amnesty upfront and waited for border security and it never came. We're not going to make that mistake. Donald Trump is not going to make that mistake … And he's talked more about going to a merit-based system, been encouraging folks who are going to be productive. He was just talking today to this senator about how the Irish who are here already, and the Irish who want to come here under, say, a new E3 programme, are all going to be productive people. Those are the types of folks we want coming into the country. We are desperate to fix one problem so that we can then fix another, and we do need to fix our legal immigration problem so it makes it easier so good, hardworking people can come here.'

There is a desire from all parties to keep this push for a change to the E3 system below the radar. Several members of Congress said they could vote for it quietly, but not if it got into the public domain. It is potentially toxic. For when Mick Mulvaney talks about the Irish being 'the types of folks' they want to see coming into the US, he will say he means hardworking and like the Irish who came before and contributed so much to the country.

However, that is interpreted by critics as the Irish being 'the types of folks' that are wanted because they are English-speaking and predominantly white.

Mulvaney says both he and the president are disappointed that they don't have some movement to announce when the taoiseach comes to visit in a few days' time.

While something small is better than nothing, the E3 visa is a very narrow provision and will only benefit a small number of Irish people who may wish to work in the US. Mick Mulvaney immediately dismisses any prospect of or plan for any wider reforms that might benefit more Irish (and all other nationalities).

'No. No, this is the one thing that's doable – in fact, we hope it's doable, and it took us a while.' He says Irish Ambassador Dan Mulhall, Fine Gael Deputy John Deasy and Fianna Fáil Senator Mark Daly were all helpful. 'There was a bunch of people – not excluding the Australians whose visas were part and parcel of this – who had to be convinced, and they were, so it was a heavy lobbying effort.'

Having an Irish American vice president and an Irish American chief of staff – both of whom care for the country of their ancestors – offers a level of influence in the highest possible office only dreamed about by other nations. Mulvaney openly admits that he lobbies for Irish affairs, as did his predecessor in the role, John Kelly. Mulvaney says, within the administration, having connections to a certain country, in this case Ireland, 'makes us more inclined to look for places where our interests are aligned'.

'I started with Irish immigration when I was on the Hill, I took Irish immigration to the Office of Management and Budget, now I've got Irish immigration here in the chief of staff's office. And the boss is, like, "Where are we on the Irish?" I'm like, "Well, boss, here's the hitch." "Oh, I forgot about that – let me call this guy [Senator Tom Cotton]." So he's trying to get it stoked back up to at least tell the taoiseach there's some progress when he gets here this week.'

But to get a bill to the president to sign, even if he is surprisingly open to it, means it will have to get through Congress again.

There was bipartisan support for this bill the first time around, in 2018, and all signs point to a similar path. During the original brief debate on the bill on the floor of the House of Representatives, several of the speakers reiterated the need for comprehensive immigration reform and that this gesture was designed to foster the relationship with one of America's 'closest allies'.[19]

Republican Congressman Peter King, from New York, is happy to have this bill pass. But, living as he does in a multicultural district, backing a narrow bill like securing the unused E3 Australian visas is one thing, but he would not be able to back

19 Congressional Record, 28 November 2018, www.congress.gov. <https://www.congress. gov/congressional-record/2018/11/28/house-section/article/H9676-1>

something wider for one ethnic grouping over another. He says he is a unique Republican in many ways because he is what he calls 'liberal on immigration'. He says he is like President Trump in that he wants a strong border, but he also wants to take care of the so-called DREAMers – those who were brought to the US as undocumented children, have grown up in the country and are essentially Americans in all but name.

'If the border is secure and people have been living here long enough, and they're not getting in trouble, we should give them a pathway to stay here. I would say the average Irish American in my district would have no problem with me voting to help DREAMers and also voting for Ireland and vice versa. I think if I just voted for Ireland and said I'm doing nothing for the others, then, yeah, I could have Hispanic groups coming after me. They wouldn't consider me Irish. They may say I'm pro-white or something like that.'

In other words, he would be called a racist – a term bandied about in the United States now in a way that it hasn't been for decades.

'The Irish don't stand out as an immigrant group the way they would have in past times. To the average American, the question of Irish immigration would be the same as any other country. They're not aware that Irish immigration is way down. If I was giving speeches haranguing Hispanic immigrants, Central American immigrants, and meanwhile said give the Irish a special class – yeah, I would be attacked for that. But working out a way to solve an anomaly where Irish immigration has almost been ended, to give them some kind of benefit, I don't think that would backfire, no.'

Pete King's colleague across the aisle, Democratic Congressman Richie Neal, agrees that any sort of comprehensive immigration reform is highly unlikely at this stage, or any time soon. He also agrees that any sort of amnesty that catered just for the

undocumented or illegal Irish nationals in the US and not any
other nationalities is 'unlikely'. Whereas King blames the lack
of progress on reform on the Democrats' lack of willingness to
engage, Neal blames the president's views on undocumented
immigrants.

Congressman Neal says what he calls 'the immigration picture'
has been 'very harmful to the Irish'. He continues, 'The
undocumented Irish are really not the controversy. It's on the south-
west border – that's where the controversy is. But we all know many
who live in the shadows back home [in his Massachusetts district].
I hope we're going to be able to normalise their status as well. But
in the broader context, that wave of European immigration that we
once got, it included a lot of Irish and that has certainly subsided.'

Neal also thinks any special deal for the Irish is unlikely because
even the most Irish of congresspeople, like himself, would find it
difficult to vote for it on the grounds of fairness, at the expense
of other ethnic groupings in their constituencies. Many of those
other groupings would not want to separate the mostly white
English-speaking Irish immigrants from the greater percentage of
migrants who do not have English as a first language and who do
not have pale skin. 'Many of the other groups had figured out that
if you let the Irish go that you'd be better off being handcuffed to
them on the way in.' This goes back to Mick Mulvaney's comment
that he and the president are looking to help Irish immigrants
because they are the right 'type' of migrant. It was not always
the case, but the modern Irish immigrant experience in the US
is different to that experienced by other nationalities who have
darker skin colours.

Richie Neal thinks the E3 visa solution makes perfect sense if
the Australians are not using the visas. But that will be the extent
of it. 'I don't think that you can do a big immigration bill until the
president relents on some of the incendiary use of words that he
has. I think that he has so charged the whole question that there's

been this sort of universal indictment of immigrants and it's an awful mistake because we really need them.'

Former congressman Joe Crowley, who was to the forefront of the charge to have the E3 visas over the line at the end of Congress sitting session in 2018, thinks it will eventually get done. He describes the effort at the end of 2018 as a 'missed opportunity'.

Crowley recognises that the Irish immigration lobby is a strong one and has been. 'John McCain, Ted Kennedy. They're gone now, but they were very much in tune with that, you know, a decade ago, but it has waned a bit, because of the lack of movement on anything, you know, it's just been …' He trails off with a sigh, looking for the right words. 'Especially with this president, but it eventually will happen. We have to modernise the law anyway; it's not working.'

Congressman Joe Kennedy III agrees that the days of doing a deal exclusively to benefit the Irish have long passed. 'I think doing something for one ethnicity is difficult. But I think that should be part of the reason why we recognise that it needs to get updated. I hear stories of folks that have been in the United States for a while from Ireland who no longer can go home to see a dying loved one, that miss out on these experiences because of our immigration system. I think we can update that, but it is obviously not just the Irish that suffer from that, or the United States that suffers from that. But it is, I think, part of the clarion call to say: Let's fix this.'

—

Back at the White House, Mick Mulvaney is always happy to weigh in on things he can do something about – like immigration – but he's not prepared to get involved in Brexit-related issues in the way his former congressional colleagues are on Capitol Hill.

'On Brexit, as of right now, our attitude is to let it play out. We've not weighed in on one side or the other. Not sure that we could help much to fix things.'

While Mick Mulvaney may consider the administration has not 'weighed in' on Brexit, the same is not true for the man he calls 'the Boss', who has said repeatedly that he thinks Brexit is a great idea. And as for not weighing in regarding British affairs, that was turned on its head in July 2019, when diplomatic cables from the British ambassador to the US Kim Darroch were leaked. The cables from the accomplished diplomat outlined his view of the Trump administration and how things were run in the White House – 'inept' and 'utterly dysfunctional'. This prompted the president to tweet that Darroch was a 'pompous fool' who he wouldn't be able to work with anymore – essentially calling for Darroch to be fired. Rather than risk a further diplomatic storm, Darroch tendered his resignation, with many shocked that the American president had essentially been able to influence the operation of the British Foreign Service in this way.

Nevertheless, what is Trump policy is not necessarily US policy.

'We still look at it as very much a European issue. We would like to see it resolved quickly. We'd like the ability to talk with the British about trade policy as soon as we possibly can. It's always stunned me that we don't have a free trade agreement with the mother country. But we don't. We have it with a bunch of other countries but not England and not the UK because they've been inside the EU, so we're very interested in talking to them as soon as we possibly can. We think they'd be easier to work with one-on-one than the EU is. But other than that we're just watching from afar.'

However, he says, they are keeping a watchful, if removed, eye on what happens around the border area.

'We do take an interest in what's happening on the back-stop just because it really is the one place where things could get ugly relatively quickly. I've explained this to the president. Re-establishing the hard border and the checkpoints could be the type of thing that has that unintended consequence of injecting violence into an otherwise economic decision. We watch it, but I don't think we are going to participate in it in any direct way yet.'

Apart from his somewhat vague ancestral connections, Mick Mulvaney has much more current ties to Ireland. A daughter studied in Alexandra College, a boarding school in South County Dublin, a few years ago. She subsequently applied for an undergraduate course and was accepted into Trinity College Dublin. She ultimately decided to go to university in the US, which Mulvaney says broke his heart, not just because it was 'pretty cool' that she could go to Trinity, but also because it was much cheaper than the American university, where she ended up. However, her period in Milltown as a teenager gave her a love for Ireland and she spent the summer and second half of 2019 on a study-abroad programme in an Irish university.

Those personal and family ties that stretch across the Atlantic between so many Irish and American people can only be enhanced, Mulvaney feels, if Brexit does indeed go ahead. He thinks it's even better news for business ties.

'If the British leave in some fashion, that leaves Ireland the only English-speaking capital in Europe. I think that relationship might actually grow even stronger, as certain components that might otherwise consider operating out of London would move to Belfast. A lot of American companies are in London because of the language, and if they still want access into the EU they can move to Dublin and that will just strengthen the business ties between the two countries.'

Once President Trump slammed Ireland as a country that was benefiting too greatly from the United States. The US launched an investigation into the dealings in March 2017, naming Ireland as the country with the fifth highest trade surplus with the United States, something he pledged to reverse in order to 'protect American workers'. Mulvaney, however, speaks favourably of the trading relationship.

'The trade between both countries is growing so that will continue. Trade between the countries is significant but it's quite nicely balanced once you look at goods and services together. So the relationship is really, really good. You folks are defending your tax treatment of the American companies, which we think is the right thing to do. So I can't think of anything bad to say about the relationship.'

So it would seem that there will be no further attacking of Ireland over low corporate tax rates and the tax bills paid by some US multinational companies with large bases in Ireland. It wasn't just President Trump who had this viewpoint: his predecessor, Barack Obama, also often used Ireland as a byword for 'tax haven'.

'No, in fact, we look to you as an example. You took what we did in the 1980s and beat us at our own game. We taught the world when we lowered our corporate rate to 35 per cent in the 1980s. It was the lowest in the world. When we changed our corporate rate in the last year, we were the second highest in the world. We taught the whole world that lower corporate tax rates do lead to economic growth. You folks are the best in the world at it. We'd love to get the corporate rate down to your level – we just can't get it through our Congress.'

Mulvaney describes the relationship between Ireland and the US as 'friendly competitors'. He admits that they've changed the tax codes to try to lure American companies back from Ireland and elsewhere, but he's guessing that Ireland will then take some other form of action to try to keep the firms in Ireland. He says he and the president call this 'healthy competition' and feel that when the economy is good everybody benefits.

Two diplomatic sticking points in that relationship had been the delay in appointing a US ambassador to Ireland and the delay in appointing a US envoy to Northern Ireland.

The ambassadorial delay was due to President Trump's first pick, Brian Burns, withdrawing himself from the running and a backlog in the Senate in confirming his second pick, Ohio man Edward Crawford. His father left Cork for America in 1925; his mother left in 1927. They met and married in the US. Ed Crawford went on to start a company called Cleveland Steel Container in the 1960s while in night school. That enterprise has grown and expanded to have a revenue of almost $2,000,000 annually. Park Ohio employs over 7,500 people. His elevation to the role of ambassador meant he retired and resigned as president of the company and from his board positions. The company made a $6 million retirement payment to him. He was eventually confirmed on 13 June 2019, and arrived in Ireland soon afterwards, officially presenting his credentials to the President of Ireland Michael D. Higgins on 1 July. After a vacancy of almost two and a half years, there was once more a US ambassador to Ireland. But there are no such Senate confirmation delays regarding the appointment of a special envoy, as was promised.

It seems that Mick Mulvaney himself may have been the front-runner to be named as envoy to Northern Ireland by Donald Trump, but with so many resignations and changes in personnel within the West Wing, the president needed a capable loyalist like Mulvaney to stay put and made him chief of staff instead. The former congressman won't confirm or deny that he was on the list, but laughs and says it would have been a great honour to be picked. He does think, though, that it is important for there to be an envoy and it is something the White House is conscious of. He's spoken frequently to Secretary of State Mike Pompeo about it and they both want to make it happen. He thinks there is a role for the US to play and is working on the president to turn that into a reality.

But words are easy, actions are harder. The wait for an envoy continues.

THE FUTURE RELATIONSHIP

WORKING FOR IRELAND

There's nowhere quite like Worth Avenue, Palm Beach, Florida – with the exception, perhaps, of Rodeo Drive, Beverly Hills, California. Worth Avenue is appropriately named, as those strolling past the shops, cafés and restaurants here are literally displaying their worth from every available limb, dripping diamonds and clunky watches and small clutch bags and handmade leather shoes – and that's just the men. Palm Beach is home to an outpost of nearly every high-end designer store you could mention, and the parking lots look like displays from an auto show.

Lunch is spent sheltering from almost year-round searing heat in shaded terraces and patios or air-conditioned dining rooms. Italian cuisine is popular, or at least the American version of Italian food.

A rule of thumb in journalism is always to dress appropriately for where you're going. It puts your subjects at ease and makes the journalist appear relatable, as well as demonstrating that you've bothered to research the subject and the environs. It demonstrates respect to the person too. If you're interviewing a head of state in their palace or parliament, you don't arrive in jeans and trainers. If you're going to a farm, you don't arrive in stilettos and a light dress. And if you're going to meet wealthy Republicans of a certain age, don't arrive in a hoodie and flip-flops.

I was dressed in what I considered appropriate attire to meet a group of Donald Trump's compatriots and friends and was having a quiet bite to eat in one of those Worth Avenue area Italian establishments. All of a sudden, I felt deeply underdressed when two tall, thin, striking women arrived on the patio dressed as if they were headed to the Oscars or a glitzy New Year's Eve party. Was this the lunch attire of choice in Palm Beach? I feared I had deeply underestimated my subjects.

The first woman, in a pale-blue feathery number, walked straight past me and to the table of three Snowbirds (the term given to those who flee the north-east-coast winter of New York for the sunny climes of Florida and then relocate when Florida becomes unbearably humid in summer). They proceeded to coo over how beautiful she looked – her dress, her shoes, her nails – and reached out to stroke the material of the gown, while asking her to do a twirl. As she turned around, the backless dress revealed her sun-scorched pink skin crisscrossed by white bikini lines. Such was the level of familiar compliments and validation that I presumed she was the daughter or granddaughter of one of the three guzzling iced teas and Sauvignon blanc. But no – the young woman produced a card and said the outfit was available from such and such a boutique on Worth Avenue if they wanted to go along and buy it or similar. She was essentially a door-to-door saleswoman, albeit a glamorous one. But here she was, plying her trade in the shade of an Italian bistro in one of the wealthiest districts in the United States.

Suddenly there was something unsavoury to this whole scene – a tableau of, not quite haves and have-nots, but rather have-lots and have-littles. This young woman, perhaps hoping to be discovered by a modelling or talent agent, was spending her day 'doing a twirl' and having her pedicure admired by much older women in the hope of directing them to a store to buy high-end designer fashions.

This blue-feathered twenty-something sashayed off to the next bistro, only to be replaced by another woman, probably in her late forties, again tall, thin and impeccably groomed, this time wearing a black and silver sequinned number. She repeated the table-side theatre with the three Snowbirds at the table beside me before also revealing the name of her boutique and then sashaying off the terrace. Outside on the pavement, she was met by a younger, more ordinary-looking woman who threw a jacket over her shoulders and gave her directions for what tables to target at the bistro next door.

This is the town in South Florida where President Donald Trump likes to spend as many weekends as possible, holed up on the golf course and high-end resort club that is Mar-a-Lago. It's a magnificent property, but it's not the only one on this little peninsula jutting into the sea, connected to West Palm Beach by a series of bridges. This is the ground zero destination for the elite and wealthy – just as on the far west coast Beverley Hills could be said to be the same. The difference being that there the elite and wealthy crowd is most likely to be Democrat or liberal. Here they are most likely to be Republican.

President Trump is a transactional sort of guy. He was when he was a businessman/developer/real estate smooth in Manhattan, and he still is now that he is in the White House. Everything is quid pro quo. Everything is measured in whether there's a deal to be made. This is instructive in relation to Ireland's approach to dealing with this president and this administration. The relationship has to be beyond emotions and culture when he has no emotional tie to Ireland and is perhaps less interested in the cultural side. Any asks that Ireland may have must be presented in terms of the meaning or worth that the president can assign to them.

Palm Beach is home to Brian Burns, a proud Irish American who has spent much of his life, like his father before him, campaigning and fundraising for Ireland and Irish causes. He's a third-generation Irish American who traces his roots to Sneem in County Kerry. Shortly after the election in November 2016, he met his 'good friend' and president-elect Donald J. Trump at the nearby Mar-a-Lago club. On that occasion, Trump asked Burns to be his ambassador to Ireland. Burns was honoured and delighted and he accepted. A photograph of Burns, Trump and his wife Melania from that night has a prominent position in his office now.

However, Burns had a stroke the following month and decided that he just wasn't well enough to take up the role. It was disappointing, he says, but he has tried to keep up his involvement in other ways. He is still involved in Irish affairs and acts as something of a go-between between the president and those in the Irish American community.

He has a long career of philanthropy and fostering Irish affairs. He funded a library in Boston College, the Burns Library, named for his father, the Honourable John J. Burns. It has the largest collection of Irish books and rare material outside Ireland. He funds a scholarship on the Irish Studies programme in Boston College, whose graduates include former Irish president Mary McAleese. He also amassed one of the largest private collections of Irish art in the world, until he auctioned most of it off in November 2018, raising $4.2 million in Sotheby's, one of the largest ever results for a private sale of Irish art.

In 1978 Brian Burns became the youngest ever director of the American Irish Foundation. That agency was founded by President John F. Kennedy and President Éamon de Valera to encourage American charitable fundraising for Irish causes. When Burns took over he soon realised that another group, the Ireland Fund, which had been founded in 1976 by Dan Rooney, the owner of the Pittsburgh Steelers, and Pittsburgh-based Irish businessman Anthony O'Reilly, was doing similar work, so he suggested, and in 1986 shepherded, a merging of the two. On St Patrick's Day 1987, at a ceremony at the White House, they merged to form the

American Ireland Fund. That has since grown and operates in twelve countries, and is known simply as the Ireland Funds.

Now in his early eighties, Brian Patrick Burns says he has a close relationship with the president, and they speak regularly by phone or in person, mostly in Palm Beach but occasionally in Washington DC.

Burns says he is a great admirer of Donald Trump, despite his own, and his family's, lifelong association with the Democratic Party. He grew up just six miles from the Kennedys in Massachusetts and the families were close friends. His father was lawyer to Joe Kennedy Senior, the patriarch of that family, and was also close with Tip O'Neill, who went on to be Speaker of the US House of Representatives. At 29, he became the youngest judge in the history of Massachusetts. Later, he was the only Irish Catholic on the board of CBS, the Columbia Broadcasting System, to this day one of the major channels in the US. In the early years of World War II, Burns's father arranged for a cable hook-up from Iveagh House, the headquarters of the Irish Department of Foreign Affairs, to enable Frank Aiken, the then minister of foreign affairs, to address the American people and to explain Ireland's position in the war.

But he says that Donald J. Trump is really a Democrat at heart too, and he can't understand the animosity that so many people in Ireland have towards him. He acknowledges that he is quite popular with many Irish Americans, but the views in Ireland are so different. He says that, while there is no Irish American voting bloc, there should be one focused on Ireland. 'In a nice, confident, quiet but very purposeful way to say we want the interests of our homeland to be considered and given some priority.

'I feel very strongly that President Trump has that feeling. He loves Ireland. He looks, as do I, at the terrible articles that are written about him in the Irish papers. Your press is even tougher than the United States. And sometimes I wonder why don't they give this new president a chance and let him go to work? He's the first president that owns lots of land and has some hopes and dreams for Ireland, and you find him a really great friend of Ireland. But, sometimes, perhaps Ireland could reach its hand out too.'

Brian Burns feels this is a dangerous game for Ireland to play. He says the relationship between Ireland and the US now, with President Trump in the Oval Office and potentially set to be there until January 2024, needs work. 'It could be a hell of a lot better, and it should be.'

Diplomacy at the highest levels can't exist without access. Early in 2019, Ireland's ambassador to the US, Dan Mulhall, and his wife Greta were invited to dinner by Brian Burns. The dining destination? Mar-a-Lago. The surprise guest they ran into while there? President Donald J. Trump. Burns may no longer be set to be the US ambassador to Ireland, but he continues to do what he can for Ireland, Irish issues and Irish people. So he purposely brought the Irish ambassador to Mar-a-Lago so there could be a long and informal meeting with the president. The president later said to Burns that he wanted to do something for the Irish on immigration and asked Burns to take a lead on getting it done. This chink of light for those campaigning spread rapidly through the Irish American community. If the presidential door was slightly ajar, many were ready to push through it.

Veteran immigration reform campaigner Niall O'Dowd says, 'That comment set off an alarm bell for us. We need to try and do something about that, and we have been trying. It's all about how you have to play the game. We're banging away now on that Trump comment. And will we be successful? We might well be.'

Brian Burns says this is the case. 'I have high hopes that President Trump and the Congress may start to loosen the bonds which have really kept Irish immigration way, way down. They could, of course, allow many more E3 visas and other ones. That whole subject should be revisited and reopened, and the gate should be opened up very wide. Let's have intercourse between the people of Ireland and the people of this country, and for the business world, I think Ireland is an ideal place. It's an entrance to the European market, it has a beautifully educated student body, and the people of Ireland are still the warmest and most gracious and welcoming of any place that I've been in my life.'

Burns says there is a problem currently in the US with an oversupply of immigrants and it has to be solved. 'But Europe, and in particular Ireland, needs a wider welcoming mat from this administration. It would be nice if Ireland looked at President Trump and opted to think, "My God, he owns property in Ireland. His son Eric and his wife, Lara, spend a lot of their time here." He has every warm motivation to Ireland.'

Ireland's current ambassador, Dan Mulhall, says that in his dealings with the president and the administration it is true: when it comes to Ireland, the president is 'always perfectly positive and pleasant'.

'He always says the right things about Ireland. He's positive about the immigration issue. He wants to be helpful on that. He understands the importance of the peace process. While he may be partial to Brexit, for other reasons, he, I think, is careful not to want to do anything or say anything that would compromise peace in Ireland. Also, he has never said anything negative about us, as such, that I'm aware of – given that he has a tendency to shoot from the hip. We haven't attracted any negative attention from him over the last two years, which is a good thing. It shows that there's a generally positive attitude towards us within the administration.'

The president's chief of staff, Mick Mulvaney, also agrees with Brian Burns's assessment that President Trump gets a raw deal in public opinion in Ireland. 'The filter that you folks see through in Europe is generally woefully biased against the president.' He blames CNN for that, because it is the largest international

English-language TV channel, and he believes they are 'out to get' the president, so people watching their coverage, he feels, don't see a fair representation of how he is. He says they don't dwell on it in the administration because none of the people abroad who protest against the president actually have votes. He says it is 'frustrating' to him on an individual level because when he goes overseas he feels he has to explain the situation. He says he spends his time saying, 'No, no, no, the president's not really like what you see on television.' But he says he gets used to it and is quick to add that politicians always complain about their treatment in the media.

Brian Burns says that President Trump is misunderstood by many in Ireland and that he has an affection for the country that is as great as that of Clinton or Kennedy or Reagan. He says he doesn't deserve the animosity with which he's treated. 'He hasn't done anything to Ireland!' he says.

'I guess until about seven years ago, he was a New York Democrat. My God, what's wrong with that for the Irish nation?'

He continues. 'Ireland's great and they ought to accept this guy. If you're building in New York, you have to make some payoffs. He was a very active Democrat.'

But despite the long connection to the Democratic Party which the president and Burns and his family have, Burns says he doesn't recognise the Democratic Party of today. Although he is a proud supporter and confidant of the president, he's not a Republican either, and given his background in Boston, he has problems with how the Republican Party treated Irish people for generations. Now he's a registered independent voter.

He explains why he thinks the Democratic Party 'has just shifted', and that there is a greater focus on LGBTQ+ issues and women's rights within the party.

'I don't know who its loyalties are to now. They think there are evidently three or four kinds of sexes, and everybody who complains has a separate following, and it's becoming almost like the Tower of

Babel. It's not how I remember the Democratic Party. They seem to have forgotten their roots in taking care of immigrants and taking care of that famous statement of James Michael Curley's, the famous mayor of Boston, "the forgotten man".'

Although not a fan of the Republican Party as a whole, he is a big fan of President Trump and knows plenty of other Irish Americans who are, for the following reasons, he says.

'I like someone who says, "This is what I'm gonna do if you elect me," and then he does it. That's President Trump. And he doesn't care. They can call him any name they want. He's getting it done. And he infuriates so many people because, no matter what they say, do or what they hurl at him, he's got a programme he's going to get to. He's going to defend the country. The Irish are very patriotic, no question about that. They were the policemen and the firemen and the people who responded so grandly on 9/11.'

Burns wraps up by saying what a close friend and nice guy Donald Trump is, and then adds, a little cryptically, 'I have done some things for him and, I'll tell you, I'm a lawyer.'

I jokingly ask whether that means he knows where the proverbial bodies are buried.

'Oh no! I don't know that; I do not know that!' he hastily adds.

—

Across the state of Florida, in almost a straight line, lies the city of Naples, another wealthy beach spot where the summer vibes last almost all year. It's become home to many Irish-born and Irish Americans. Plenty of Irish names recognisable from the Celtic Tiger era and beyond have magnificent homes here. It's hot and flat and the must-have accessory is a swimming pool. At one such fine home, Sunday brunch is under way with some Irish-born and Irish American friends. The men and women gathered are a mix of ages, backgrounds and political persuasions.

Naples is not a place that springs to mind immediately when thinking of Irish power bases in the United States. But it's like many places in this country, where there are pockets of Irish people in obvious, and not so obvious, locations. Some want to connect to back home, others don't. Some embrace the Irish American view of Ireland, and others don't. But each in their own way act as ambassadors for Ireland and can have an influence on the local community. While the outside view of the Irish community in the US is that the Irish align with the Democratic Party, it's more a case that the Democratic Party has been more organised and better at rallying Irish people in bigger cities and in traditional power bases. But stepping outside those green spots is often where the Irish and Irish Americans who are Republicans and who support the current president can be found. And plenty of them are in Florida.

Bridget and Fintan McIntyre are two of the guests at this casual Sunday brunch in a luxurious home in Naples. They are from Ireland and still spend several months a year there. The rest of the time they live in a permanent well-kept trailer park in Naples. It's a settled community and they say most of their neighbours are Snowbirds.

Bridget and Fintan are registered Democratic voters but say they actually vote independent. They proudly cast their votes in 2016 for Donald Trump and say they plan to do so again in 2020.

They've had a vote in the US since 1980 and have regularly voted Democratic. Fintan even canvassed for Democratic candidates when they lived in San Francisco and he was the precinct captain for the party.

They believe at one point there wasn't a huge difference between the parties – they were close on social issues like abortion, and there wasn't the level of divisiveness they see currently. Bridget says that now 'the centre is caving in; the middle classes are under pressure'. Fintan says a big issue for him is how the media attacks Donald Trump. He describes it as 'unrelenting' and it only makes him deepen his support for the president.

The McIntyres lived in New York for a time and Fintan worked for a law firm that specialised in reducing real estate taxes. He says he had some dealings with Donald Trump when he was one of the city's big developers. He tells the story of how one day he ended up speaking with Trump directly on the phone.

'He lifted the phone, and when he heard my Irish accent, he said, "Why don't you come up and work for me? I've got Irish guys working here – you know, all these security and these doormen and maintenance people, all Irish."'

While Bridget and Fintan would like to see immigration reform, they're not against illegal immigrants in the way the president purports to be. Bridget is a nurse in a senior citizens' facility, and she says that where she works, 'if all the illegal aliens left there, everybody would die. All the old people would die.'

They voted for Bill Clinton in 1996 and say they actually liked Hillary Clinton for a while, but they felt she was getting pulled too far to the left, becoming too liberal. On that note, they feel Ireland has become too liberal and is echoing some of what's happening in the US.

Bridget says that, although she is a nurse and at one stage worked as a midwife in England in a hospital where abortions were performed, the Democratic Party's position on abortion and permitting later stage abortions is not something that sits well with her.

Fintan says the Irish community in America often discusses how 'you go back to Ireland now, they don't even want crucifixes on the hospital walls. And, like, all these Muslims being allowed in.' He doesn't expand further on what he means by that.

Terry McAuliffe, the former governor of Virginia and chairman of the DNC, understands their motivations.

'Trump was able to pull off a significant amount of Irish voters with a false promise. I'm going to get us out of deficits, I'm going to give you tax cuts, none of it's come true. But I think a lot of people, not only Irish, but a lot of Americans actually believed it and now we have a little bit of Irish remorse today in the country. Rightfully so.'

But McAuliffe says all these issues are contributing to a further breaking down of any possible notion of an Irish voting bloc.

'I think what Trump has done on immigration has been a disgrace and I think that's offended the Irish and I think it's really offended many of those individuals who may have left to go on over and vote for him who literally believed the blarney he had. And I say that – blarney – politely.'

New York Republican Congressman Peter King believes those who are over 40 are more 'socially conservative' and those who are younger are shifting away from the traditional right-of-centre position. He gives his own example of what he includes in his newsletter to voters.

'When I started out in '92 running for Congress, you sent out mail so you could target different groups. Leaving the Irish issue itself aside [by this he means peace in Northern Ireland], if I was talking about a strong defence and strongly defending the police, I was sending it to Irish Republicans and Democrats, I would get a good response from the Irish Democrats. Today, I don't get that good a response from the Irish Democrats because most of them have become Republicans.' So he already has their votes in the bag.

He continues, 'They're the ones we would call Reagan Democrats. They were very much Irish Democrats at the time. I would say if you wanted to categorise the Irish voter over here, certainly in people over the age of 40, it would be blue-collar Republican, in that they believe in strong defence. They believe in standing with the police. They believe in law and order. They basically believe in hard work. They don't want any freeloaders around. On the other hand, they also realise that people do need a break. So, they're not hard-assed. They're not hard-nosed. They're sort of conservatives with a realisation that working guys have got to get a break.'

This sounds like a list of attributes of a 2016 Donald Trump voter. Strong on defence, law and order, fair play and working hard to get ahead in life. The kind of voter that Pete King has in his New York district, but also the type of voter that flipped from the Democrats to vote Republican in Pennsylvania, Michigan and Wisconsin.

'They're not anti-union. They just have, what they see as a more realistic view that, hey, if you're not going to work, you shouldn't be getting welfare, you shouldn't be getting relief. But on the other hand, if you're really going through a tough time, we'll give you something.'

Pete King is worried that he's oversimplifying what he sees as the Irish American Republican voter in his district. 'It's not a strict ideological view that if a guy is out of work, he's a bad guy.' But it is, perhaps, a view that an immigrant class who has worked hard to pull themselves out of poverty feels others should do the same thing.

For Fintan and Bridget McIntyre in Florida, they would like to see a plan to legalise and offer a pathway to citizenship for the undocumented Irish, but they don't believe anyone should 'skip the queue' for visas, and although not totally happy with the immigration picture, they do still plan to vote to re-elect President Trump in 2020.

—

Back at brunch in the Naples sunshine, over delicious food and wine, the conversation moves around to politics. They tend not to wade into this subject matter much as the discussion can get heated. Bridget and Fintan have strong views. But Ross and Noreen Edlund, who are hosting brunch today, are Democrats and viscerally opposed to what President Trump stands for and what he is doing. Noreen is from Carrick-on-Suir in County

Waterford and her husband, Ross, is Norwegian American from Minnesota. They run a chain of breakfast and lunch diners across Florida and, for them, and particularly in a state like Florida, the economy is fuelled by an immigrant workforce. Some have the right paperwork, and some don't.

Ena Gleeson from County Mayo and her husband Terry Gleeson are card-carrying Republicans. Mary Doherty, who is American-born to Irish parents from Mayo who settled in New York, says she's also a registered Republican because it's such a red town that often there is no Democrat running for election and, the way the system works, if you're not registered as a Republican you won't be allowed to vote in Republican primary elections. Ross says 68 per cent of the town voted for Donald Trump in the 2016 election, so if you're a Democrat in Naples, 'you're a pariah', he says.

And so, with the issue of immigrants, illegal or otherwise, raised, a good-natured but intense row between the old friends breaks out. Ross says he's not saying just anybody should be allowed in, 'but they can't slam the door, like Trump's trying to do'.

Ena responds, 'You're saying it should be an open door. Anybody who knocks and says, "I want to come in – can you let me in?"'

'No, I don't think it should be open door.'

'So who should be allowed in?'

'I think we should have people that we need.'

'Like, what? Do we need people coming over the border and screaming and saying, "I want asylum"?'

'They're not. Immigrants are coming, and when they walk in the door, the first thing they get is a job.'

'They're not.'

Ross continues, somewhat exasperated, 'They are. They're not on welfare roll books. They don't want to bring their stuff from their country and become enclaved. They do want to be American. I've seen it personally for the last 45 years. Because I've worked with these people. And they're brilliant people. And I love them.

And they have family values, good values, and we should embrace them – instead we are spurning them. And it breaks my heart.'

Nevertheless, Bridget says that in the senior citizens' home where she works, some lifelong Republican voters, aged 90 to a hundred, actually changed their allegiance in 2016 and voted Democrat because they didn't like Donald Trump, his behaviour and what he represented. And even though she voted for Trump, she knows he wasn't a good candidate. But he was 'the lesser of the evils'.

Some of her patients also loved him. One man, who was 92, said to her, 'Trump was elected as a disrupter to the leftward shift in America. America was going so bad and was on such the wrong track that we needed a junkyard dog to snarl back.'

Ena Gleeson also voted for Trump and understands that the president can be 'boorish and arrogant and mean and like a bulldog to get his agenda across. It's not against the law. But it is offending people's senses.' But, she says, his policies are more important. 'To abort a baby at full term. Destroying the middle class and thinking that's fine. I'd be a lot more upset about things like that than I would be about somebody cursing and swearing.'

Her husband, Terry, knows that many people at home in Ireland don't understand his support for Donald Trump. 'Trump's personality and behaviour and mannerisms are calculated. They are exactly the kind of thing that go down like a lead balloon in Ireland particularly. We don't like people who boast and who are, you know, the "typical ugly Yank", et cetera. But the criticism of him in Ireland is over the top.'

———

With ever-evolving newer immigrant communities becoming more and more engaged, the key to the future of the Irish community, in the face of declining immigration, may be through increased activism. If Joe Crowley thinks part of the reason he lost his seat in New York was not because the Irish ceased to vote, but because other communities became more involved, the solution then may be in increasing involvement.

The vice president of the Irish American Democrats organisation John McCarthy says, 'in swing states like Ohio and Pennsylvania, and then in states that have big export capacity, like New Jersey and New York, that can get volunteers or donors out to other states, there's proud Irish Americans and Irish American festivals that will get tens of thousands of people. So my argument to the party has always been if people are willing to sit and say, "I'm proudly this," meet them where they are. Throw up a voting registration booth at the festival or something like that.'

Which is exactly what Sandi Maguire did at the North Texas Irish Festival. She's had a DNA test that shows she is just 20 per cent Irish, but says she feels 100 per cent Irish. Sandi and her husband, John, have been volunteering at the North Texas Irish Festival for several years now. It's staged over three days in a park on the edge of Dallas every March. About 70,000 people come and it is a veritable banquet of Irish song, dance, music, food and culture, sprinkled with a heavy serving of bright green feathers, green Mardi Gras beads, green glitter hats and kiss-me-I'm-Irish T-shirt slogans. The Irish consul general is a revered guest and there are stalls selling every conceivable product as well as several promoting Irish societies like the Ancient Order of Hibernians, the local sports clubs, Irish language clubs, Irish dancing and church clubs.

The takeaway food of choice is Irish stew and homemade soda bread. The drink is Irish stout – even in the hot Texas sunshine. All the attendees are wearing green, some proudly decked out in GAA jerseys, rugby jerseys and even a few World Cup Italia '90 T-shirts.

For the first year ever, a voter registration drive is under way. It's not that they think there is an under-represented Irish voting bloc in Dallas – rather, it's a case of keeping everyone engaged and making sure newer generations of Irish and Irish Americans remain involved with politics.

This was all Sandi's idea. She's a deputy volunteer voter registrar so she can register those who are qualified herself. She belongs to a non-partisan group called Clean Elections Texas.

'It has bothered me in this country that politicians really seem to try to keep people from voting – at least, the people they don't think would vote for them. So they try to take away their right to vote and I just think that's wrong. They don't have to vote the way I would want them to. But every citizen has a right to vote. And I believe in trying to protect that.'

Texas is now a strong red state – it wasn't always, and the bigger cities still lean Democrat, although the state hasn't voted for a Democratic presidential candidate since Jimmy Carter in 1976. Although it's perceived as a Republican stronghold, due largely to some conservative views around gun control and abortion, a Gallup poll in 2018 showed that, while urban areas tend to be 70 per cent Democratic and rural areas 80 per cent Republican, the state-wide variance is only 3 per cent. A low turnout from Texan Hispanics is all that's thought to be keeping Texas from becoming a swing state in national elections. Given the size of the state, a swinging Texas could make future presidential elections very interesting indeed, as winning its 38 electoral college votes would set any candidate firmly on the path to the 270 needed to win the White House.

This, perhaps, explains the desire of an upwardly ambitious Democratic politician like Beto O'Rourke to align himself with the large Hispanic population rather than his ancestral white Irish background – his full name is Robert: Beto is just a nickname.

There aren't too many attendees at the North Texas Irish Festival who are Irish-born. It's run by Sheri Bush, who has only been to Ireland once. Although she runs the festival, she doesn't identify as Irish American because it's only a piece of her genetic makeup – she's a mix of Irish, Scots, English and Native American. She says she's embraced the culture but doesn't think she has the right to take on the label. Nevertheless, she feels part of the tribe and dedicates much of the year to this Irish takeover of this part of Dallas. She only got into the festival, she says, when she met her husband, who plays Irish music, and now, 25 years later, she is the president and musical director.

While there are bigger festivals, like the Milwaukee Irish Fest in Wisconsin and the Celtic Classic in Bethlehem, Pennsylvania, this is one of the bigger annual Irish festivals in the United States. It spreads Irish culture and fun, but Sheri says a good portion of the attendees don't have any Irish heritage at all.

'A lot do, but a lot of them come because it's fun. We've introduced a lot of people to this music and, I don't care who you are, you can't not like Irish music. The rhythms and the music and the songs, they grab you and they'll take you down that rabbit hole with them. So we have a lot of people that come out that may have no connection at all. But they're developing a connection out here. And we always tell everybody in March, if you shake anybody's family tree hard enough, eventually an Irishman's going to fall out. So, come on, you can be Irish with us!'

The value of this sort of soft power, a quiet influence, is one that can't be accurately measured by diplomats or governments. It's a people-led power that gives emotional feel-good connections where perhaps no physical or biological connections truly exist. And in a future where restricted immigration may reduce the number of Irish-born immigrants, it is a way of keeping the power and influence alive. In North Texas, it's how they reach out to those without obvious Irish ancestry, but the same principle can be used on Capitol Hill, says Republican strategist Patrick Lyden.

'There's a lot of different constituencies playing for power on Capitol Hill and with the administration. Irish events usually have a fun aspect. You know you're going to get some good music. You're going to potentially have a good reception if you show up at an Irish event. And I think that helps because if you have a list of ten things that you have to do, you're going to search out what's going to be a little bit more interesting and what's going to be a little bit more fun. I don't think it's a bad thing to play up that angle because you get people in the seats or fill your venue and then you can hit them with the message and more of the impactful thing that you're trying to push.'

So the question facing Irish diplomats and government officials, and all those who benefit from the position that Ireland and Irish people enjoy in US life, is how to keep those policy issues, that 'message' or 'impactful thing' that is trying to be communicated, on the agenda.

For all his love of Ireland, for all his Irish blood, for all his work behind the scenes through the Troubles and into the Good Friday Agreement, Republican Congressman Pete King has some concerns about the future relationship between Ireland and the US. He goes as far as to identify a growing 'anti-American attitude' in Ireland.

He says the relationship is 'basically good', but he has 'real issues'. 'There's a developing anti-Israel attitude. In many ways, it's almost like an anti-American attitude in Ireland when there are people picketing Shannon Airport when the troops are landing to refuel.'

He's referring to the vocal position Ireland has taken on a global stage in calling for the rights for the Palestinian people and in condemning the decision of President Trump to move the US embassy to Jerusalem and recognise it as the capital of Israel.

And his observation in a time of Brexit and the tussle of whether Dublin aligns with Brussels, Boston or, indeed, Bournemouth, he feels Irish people are becoming more European than Irish, especially the younger generation.

Although there may be some cracks appearing, and perhaps that is a sign of Ireland maturing as a democracy in its own right, approaching a hundred years of independence, Congressman King does believe the relationship between the US and Ireland will stay closer than the relationship between the United States and almost any other country. 'You see that with tourism. You see it with the St Patrick's Day parade.'

And for him it goes beyond the green beer and the hats. 'All that stuff is just dopey. It's the reason for the celebration that matters. People enjoy enjoying themselves with the Irish. I don't mean the stage Irish. It's also the feeling that when things get tough the Irish are going to be there. So, it's also that business relationship between the US and Ireland. And even though there are some Irish political leaders, governmental leaders who are very critical of US policy, it never seems to make it into the personal stage. So maybe the Irish get away with more!'

He's understandably sensitive about remarks made by Irish politicians – leadership and the rank and file – which have been incredibly critical of President Trump. Comments that began during the campaign trail intensified on the night of his election and continued on into his presidency. Remarks from TDs and senators, including from the minister who would later be taoiseach, Leo Varadkar, who said in early 2017 that he would not invite President Trump to visit Ireland. Much was made of this when he assumed the office of taoiseach a short time later and ultimately went on to greet the man he had criticised in his own home on the most holy of Irish American days, St Patrick's Day, and to issue an invitation to visit Ireland.

These views further highlight the professional political alignment in Ireland with the Democratic Party over the Republican Party, which is perhaps out of step with the lived experience of Irish Americans and the Irish in America. As previously noted, there are plenty of Irish surnames in the current administration, as there were in the Obama administration. However, at the 2016 selection conventions, those of us working were falling over Irish politicians at the Democratic convention, but it was like searching for a needle in a haystack to find any 'observing' at the Republican convention.

Although Congressman Pete King nominally believes in President Donald Trump's position that the immigration regime needs to be tightened up, he does concede that the lack of 'fresh Irish blood' coming to the US could diminish the influence levels into the future. But he feels the ability to travel quickly and cheaply between Ireland and the US more than makes up for that.

'Most people my age and their parents never had an opportunity to go to Ireland, and their older immigrant parents or grandparents who came here generally didn't have the money to go back. Now, people go to Ireland the same as they go to Atlantic City. It's common that people with relatives over there come back and forth. I have cousins from Limerick city who come up here, you know, for weekend shopping excursions. They wouldn't go to Tipperary before, and now they're flying over to New York as if it's nothing. So I would say so much more contact will partially make up for the lack of immigration.'

Contact is one thing, but former US ambassador to Ireland Kevin O'Malley thinks the future relationship will be commerce- and trade-based. The immigration and cultural and emotional ties will be there, but it will broaden out.

'In the old days, it was Michael and Elizabeth O'Malley coming over with seven kids and twenty dollars in the bottom of a boat with nothing more than hope. That's all they had … So that part's gone, and even if immigrants were to arrive by Aer Lingus in business class, that part probably isn't going to happen again. However, the switch has been in the Irish commerce that has come to the United States, and it looks different than the penniless immigrant that came with just a teapot. But it's still powerful, and it still has an effect, and I think the more Irish commerce grows in the US, that connection will be helpful.'

Irish firms creating employment in areas that need it in the United States will create goodwill and good feeling towards Irish business and, by extension, Ireland. For example, in 2015, Greencore opened a sandwich-making facility in Rhode Island, creating 400 jobs locally. At the time, the governor of Rhode Island, Gina Raimondo, described it like this: 'We in Rhode Island have to be laser-focused on creating middle class jobs. I am going to move heaven and earth to ensure that companies like yours continue to be successful.' In addition to creating hundreds of new American jobs, the 9,900m² factory also uses 900 tonnes of Irish cheese each year – a positive employment story on both sides of the Atlantic.

O'Malley says that even on the contentious issue of tax policy Ireland and the US have shown that the relationship can soldier on. He points to the large amount of foreign direct investment from the US to Ireland as a connection that cannot easily be unravelled.

'I know that there was a lot of gloom and doom about how hanging the tax code would affect American investment in Ireland. I never thought for one second that would make any difference. I think that Americans invest in Ireland for a whole bunch of reasons, and the tax was one of the five, but not the major one.'

He sees the future relationship being based on a more even footing than it was in the past, with Ireland investing in the US and the US continuing to invest in Ireland. But although, as he points out, the 'products and the services' are good, the deeper tribal, emotional, primal connection will always live on.

'We get one another. We, in America, get the Irish, and we like getting the Irish. The Irish understand America and, apparently, like America. So we get one another, and I think that is the reason that this will continue to go. I wish I could make a really intellectual-sounding economic argument and come up with a big name and patent it, but really, the case is, at the cellular level, we get one another, and we'll always be together as a result of that, as long as we keep nurturing the deal.'

Ireland's current ambassador to the United States, Dan Mulhall, points out that the relationship will change into the future, and has to change, and he too argues that it will become more focused on economics and education. But he agrees with O'Malley that the cellular-level tribal connection will always be at the centre of the relationship.

'The affinity and the sort of family connection between Ireland and the United States, the positive feeling towards us based on our history and heritage, would continue to be a core of the relationship. But I do think that is being augmented now by other elements, which will be more and more important in the future. I think the economic relationship now is genuinely significant, even in American terms. I do think, also, that the educational links are quite meaningful now. Wherever I go I come across links between universities in America and Ireland, either with a formal programme with an actual university, a degree programme, a shared programme, or arrangements for students to go for a semester or for shorter periods to Ireland. That's really thriving.'

Democratic congressman and chair of the Friends of Ireland caucus on Capitol Hill, Richie Neal, also feels that economic ties are the way ahead, but as an underpinning to the human relationship that already exists. 'I think that the bonds are unshakeable. But there needs to be a reminder of the role that America played in terms of direct investment in Ireland, even during the worst moments of the recession.'

He says US foreign direct investment in the aftermath of the collapse of the Celtic Tiger sent important signals internationally, as well as helping out with jobs and boosting the economy in the short term.

Congressman Neal says it was important for people to see that America saw Ireland as 'a good investment'. That fuelled a second upside for Ireland – positive contagion. 'America is always going to redeem that dollar. There's never a question. You can put that dollar out anywhere that you travel. People see that dollar and say, "Yep, they're going to make good on it."' So if Ireland was good enough for America, it was good enough for others.

Former Republican congressman Jim Walsh says that, for him, the main way to preserve Irish influence in the US, and to keep the relationship alive, is to keep it malleable, so that Irish America can step in to help or stay quietly in the background, as the case may be.

The relationship should be 'sort of as needed', he says. Ireland should not rely too heavily on American politicians but do enough to foster connections that can be called upon. From a politician's perspective, he says, 'you deal with what's in front of you, and if you have a crisis, you have to deal with it. If you don't have a crisis here, you might have a crisis over there, so you deal with that crisis. You've only got so much bandwidth. When things are good there's no real need. You take your energy and your time and you spend it elsewhere.'

When called upon, Irish America and the wider US political scene stand ready to use that 'bandwidth' for Ireland, he says. 'We shine a bright light on what's going on. I don't say this with any sense of self-importance, but if the US is interested in a place in the world, the rest of the world's interested too.'

CONCLUSION

STARS, STRIPES AND SHAMROCK

As Yeats might have put it, Irish America as a cohesive political voting bloc is dead and gone, it's with O'Leary in the grave.

Is Irish power and influence overall really on the wane in the US? That is one question that only historians will truly be able to judge with the passage of time, but it is certainly undergoing a period of 'chassis', as Sean O'Casey might describe it. The population is mature, not in age but in tenure. The issues that unified earlier Irish immigrants are not as pronounced. The need to send money to the starving in Ireland, to send reinforcements to fight against the British Crown, to send help for the peace process has passed. Ireland is a wealthy country now – Irish citizens want to move to the US for adventure and opportunity, not to find their fortune as they once did.

The challenge for Irish diplomats and government officials is how to future-proof the relationship in an era when many other immigrant communities are fighting for attention and the US itself is going through a major period of political upheaval.

The relationship between Ireland and the United States, for many of those at the forefront of it, is based on emotion and pride. It is a cellular-level, tribal connection that goes deep to the heart of their identity. That bond exists, for many, regardless of whether they can actually trace the lineage that gives them the blood, no matter how slight the amount.

But if the relationship is based on emotion and pride, how meaningful and resilient can Irish power and influence in American politics be into the future? How do you keep it alive when you move further and further from Irish-born immigrants? The first-hand accounts of famine and rebellion and war and republicanism and the fight for freedom that were handed down through generations have less sticking quality when Ireland is now progressive, liberal, multicultural and, in a global sense, wealthy. Affordable air travel or a perusal on Google Street View removes the romance and mystique of 'the old country', once left behind, never to be seen again.

Emotion, the pursuit of a good time and the power of St Patrick's Day will continue to keep the door ajar. They will continue to permit diplomats, politicians, businesspeople and any other Irish person who needs access in America to get their foot in the door. Building a relationship is never guaranteed beyond that, and that is where careful cultivation and nourishment is required. Once the relationship is solid, actually making use of it is an entirely different consideration. There will never be a shortage of takers for invites to parties at the Irish embassy, and, with the code of polite political networking in Washington DC, that will lead to a willingness to meet the travelling taoiseach, tánaiste or minister for a quiet tête-à-tête on Capitol Hill. The key is keeping it meaningful, guaranteeing that it does not fade to 'just a photo op'. Successive tours of Irish diplomats in the US have worked hard at that, but the effort needs to be broader.

Director of the Northern Ireland Bureau Norman Houston has a cautionary tale. Since 2007, when President George W. Bush was in office, a diplomatic solution has evolved to take account of Northern Irish politicians who travel to Washington DC. As they are not heads of state, they are not entitled to a formal audience as the British prime minister or taoiseach would be. So an arrangement was reached whereby the first and deputy first ministers of the Northern Ireland Assembly would meet the US vice president and then the US president would do what's known as a 'drop by', as in he would literally drop into the room to meet the travelling representatives. But as the Northern Ireland institutions haven't sat since 2016, the meeting has not taken place since then.

Obviously, there has been a change in the US administration, and there is now real concern that a meeting that was so firmly on the agenda has now dropped off it and will be difficult to resurrect whenever the Northern Ireland Assemblies are restored. 'It's not guaranteed. I'd hate to lose it, but it's not automatic. It's just very difficult without a minister coming across. Ministers mean access.'

—

Arguably, the connections now live on through business and education ties. If an Irish company opens a new factory in Pittsburgh or Portsmouth, Ohio, or any part of the US where employment is low and hope even lower, a kind, generous Irish employer will garner untold amounts of soft power. Equally, a student who gets the chance to do a gap year in Ireland, or study for a master's degree there, will fondly remember that experience and keep it with them into their years as potentially a captain of US industry. Economics and education offer a connection to those who do not have an ancestral link.

Traditional Irish immigrant pastoral care centres are reporting that their services are now in demand from other ethnicities facing the sorts of problems that Irish immigrants once faced – work permits, tenancy agreements, union membership and the basics of life in a foreign country. But a stalling in Irish immigration, especially at non-professional level, means there simply isn't the same rate of requirement for these services as there once was. Yet perhaps this is soft power in its own way. A new generation of largely Hispanic and Asian immigrants may remember fondly the helpful Irish person who sorted out their paperwork and perhaps some reward or connection will follow in the future as a result.

It is extremely difficult to come to the US without a job offer any more except for those who are extremely gifted, or, as the US Citizenship and Immigration Services calls them, 'Aliens with Extraordinary Ability'. More recent Irish immigrants now arrive with job offers and attractive professional salaries. These newer immigrants often do not want to be part of an Irish ex-pat community, or those who do choose to form business networking alliances, rather than groups that help with finding affordable tenancies or immigration lawyers.

One such newer organisation is the Irish Network – a network offering professional and personal development, focusing more on supporting immigrants in the US than fundraising for causes in Ireland.

The Ireland–US relationship is no longer about Ireland the taker and America the giver; the connection is more equal and operates in both directions. The Irish Network embraces that. Stella O'Leary, the veteran organiser of the Irish American Democrats, says that she feels the Irish Network, with its non-partisan approach, will develop 'a more intellectual, new group of Irish'. They can hold events at the Irish embassy and consulates around the US because they are not representing one party or another: they are about Ireland and being Irish.

'That might be the future,' she says. 'That might be how Ireland's influence develops which would then mean the communication will be at a different level. Whether that network will still influence politics? I'd say absolutely, yes. I think they could be a powerful influence in the future – and yet they've been quiet enough so far, setting a nice non-threatening tone.'

A challenge, too, is to keep younger Irish Americans interested in Ireland. They may not want anything to do with the old tradition of sending charitable donations to a country that in many parts of US life still has a perceived (yet outdated) label of social conservatism. Yet this is the vision that older Irish immigrants are looking for. Maybe they struggle to become acquainted with a modern country – the true Ireland of today that has nothing to do with sheep in the parlour and where Aran hand-knitted jumpers are more likely to be worn by hipsters or reimagined into twenty-first-century fashions. Go to virtually any community event in Irish America and, with a few exceptions, the age profile will be mature, comprising a mostly grey-haired community. This poses a major problem for the future.

The post-Brexit world will be an interesting one for the Irish–American relationship, and conversely for the Irish–British relationship. Brexit has demonstrated the level of influence that Ireland has and can have within the European Union. The prospect of losing our one-time ally has forced the forging of arguably greater alliances with France and Germany and other small European nations, particularly those who stand to be most affected by the fallout.

As is evidenced by much commentary from British politicians and in the British media, some of it harking back to the bad old days of anti-Irish rhetoric, the British have perhaps been surprised that little Ireland doesn't need them any more and has greater influence in Brussels than the UK does.

That's the situation off the east coast. Off Ireland's west coast, one of the most powerful nations on the planet is coming out to bat for it too. Although the president of the United States might be tweeting and talking about the wonders of Brexit and the excitement of a US–UK trade deal, that prospect looks to be a long way off, as strong, long-serving Irish American congresspeople – critically, from both parties – are pledging to lay down their verbal and political swords to protect Ireland in any trade negotiations.

In fact, given that even holding meaningful negotiations on a trade deal can only begin in earnest when the UK has left the European Union, one can't help but wonder if the point of making statements about it and tabling resolutions now is actually to create a sort of uncertainty that may give the British government and people a reason to think twice.

A US–UK trade deal is not straightforward and won't automatically follow Brexit, and will be cognisant of other factors like responsibilities on the island of Ireland. Just as in the early 1990s when Irish Americans used British media attention on the Gerry Adams visa issue to bring pressure to bear. That chapter of history leading up to the Good Friday Agreement, when the president of the United States put the might of his office and government behind the wishes of the people on the island of Ireland, arguably made the British sit up and realise that they couldn't have unanswered powers of persuasion over the Irish. Now it would appear that something similar is happening. Perhaps all of these soundings are contributing to a thought process that is forcing a rethink rather than a confirmation of hard-line positions.

Now that is soft power.

The question that Irish diplomats and government officials must now ponder is how to protect this powerful transatlantic relationship. It is clearly cyclical; a low-maintenance friendship that can be called into action when needed – Brexit being a case in point – but at other times, can be left fallow. Does it suit Ireland to have it that way? Not everything the US does is in keeping with the values and foreign policy priorities of this island, so it is convenient at times to be able to ignore some US positions, to not have an automatic requirement to agree.

While there can be a speedy jump to claim anyone powerful in the US as Irish, no matter how distant the connection (step forward Barack O'Bama), there is a slowness to welcome those whose belief systems perhaps don't match up with the majority in a contemporary Ireland (step forward Steve Bannon). How does Ireland preserve alliances on the Hill and in presidential administrations if some of the personnel are viewed with distaste, or distrust?

Ireland's greatest calling card is the St Patrick's Day festival. Ireland's greatest brands are the rich cultural history, contemporary and traditional, of our poets, writers, musicians and artists. But there are great assets too in our entrepreneurs and businesspeople who can create jobs in the US and bring a positive view of Ireland into areas that, perhaps, are off the beaten track.

Nobody involved in Irish American politics will argue that there is a cohesive, monolithic Irish voting bloc, a substantive Irish vote or any major national benefit to running as the Irish Guy or the Irish Girl any more – it is a sign of a mature and assimilated population that these ethnic enclaves have gone. However, the Irish have a great tradition in the United States of being involved in public service and running for public office and that should be encouraged. Only when Irish Americans continue to hold positions of power can Ireland continue to enjoy the level of access that it currently has. The need to keep communities active, engaged and organised around local candidates is perhaps a matter for Irish America rather than for Ireland.

Mayor Ray Flynn, a long-time graduate of the Boston voting machine, argues that the concept of an Irish vote is dead, but so are the efforts to organise the Irish voters. This is something the apparatchiks in the Republican and Democratic parties agree on at national level. There isn't a cohesive plan or taskforce of people organising Irish voters in either party. Chief speechwriter to President Barack Obama, Cody Keenan, who worked on both of Obama's presidential campaigns, remembers no targeted effort from the campaign or the Democratic headquarters to get the Irish vote onside. Sean Spicer, director of communications for the Republican Party during the 2016 election and later press secretary for President Trump, says their party also never made a concerted effort to target and court Irish American voters nationally. Local-level party faithful may have done so, but it was not something the candidate or party headquarters were ever aware of.

Another reason for the lack of an Irish voting bloc in the US any more is because Irish Americans vote for both parties. They are Republican and Democrat and increasingly independent. The community itself reflects a trend happening across the United States whereby the party system is shrinking when it comes to presidential declarations, to be replaced by voters who identify as independent and wish to choose a candidate based on who is before them on the ballot paper, not just on what party they represent. Irish American voters are now American voters like any others, particularly once you move beyond small local or community-level elections.

As domestic politics in the United States moves to be ever-more polarised and hyper-partisan, a neutral stance is a safer position for Ireland to take, not aligned perhaps as per tradition with Democratic politics, but more centred. Irish American Republicans do feel isolated at times. Irish domestic politics and viewpoints, but not the country's moral compass or conscience, perhaps need to be left somewhere over the Atlantic when it comes to exercising full influence and maximising the power that can be attained in the United States.

Close friend of President Trump and one-time candidate for US ambassador to Ireland Brian Burns cautions against letting politics get in the way of the relationship. 'Ireland has a great opportunity. I want to have it throw open its arms and really reach to America, and America reach to Ireland. Sometimes I get exasperated by the politics, sort of flailing around and so forth. I almost want to say to politicians, "Get out of the way, we've got business to do."'

President Trump has often said that he is in favour of the UK leaving the European Union. He does not see the strategic benefit for countries in being members of a wider body like the EU. Burns shares that feeling. Both he and the president are anxious for the UK to be outside the EU so they can do a trade deal that would be 'very, very attractive'.

'The European Union is sort of an exasperating place for me,' Burns says. 'They issue all sorts of edicts and so forth. We have so many rules in the United States we barely get clearance there, and then somebody from Brussels decides to show their authority.'

However, with the polarisation in the United States and its own somewhat unpredictable foreign policy, Ireland needs to cultivate its relationship with the European Union as strongly as it does with the United States.

The twenty-year Republican congressman Jim Walsh also sees the splintering of Ireland from Irish America and towards Europe, but does not think the two relationships are mutually exclusive.

'Ireland, more and more, is a European country, and more and more they will look to the EU to help solve their problems than the US. But the relationship between the US and Ireland will always be very deep and strong. Look how strenuously Ireland has held on to its European citizenship, on to the European Union and how important it's become in the Brexit process. The EU has really stood with Ireland. It's stood with its partner state. Whereas President Trump has been, "Get out of there, get out of the European Union – what are you doing there?" Those are not mixed signals Ireland is getting from the White House on Brexit. Maybe that's a temporary thing, but it doesn't help.'

Of intergovernmental relations, Walsh says he thinks the White House thinks they have great relationships with Ireland but adds that he's not sure that the Irish government feels exactly the same way. But, he adds, 'on a people-to-people level the Irish–US relationship is exquisite.'

Former US ambassador to Ireland Kevin O'Malley looks to the words that President Obama said when he was on a brief visit to

Ireland during his time in office. 'He said that, Ireland and the US, our best days are ahead of us, and I truly do believe that. Having witnessed first-hand the commercial and familial relationship now between the US and Ireland, I believe that it will continue to prosper and grow. And it'll look different because it'll be a lot more commercial and a lot less personal, but it will always be a pleasant, successful venture.'

Democratic strategist John McCarthy has no fear of the political influence waning. His family has been in the United States for three hundred years. His parents have never been to Ireland, yet they raised him as a staunchly proud Irish American.

'Irish America is something totally different. You don't need to have people come from Ireland to have Irish America. Irish Americans will just keep being Irish Americans. I hate that cliché about everyone's Irish on St Patrick's Day, but the brand of Irish America is so, so strong that people want to be a part of it. I actually think it has nothing to do with immigration. It's too strong to go away.'

Many of the Irish American politicians featured here have mentioned with almost crusade-like fervour the notion of 'social justice' as a compelling reason for getting involved in politics. For some, it's because they've grown up on stories handed down from the Famine times; for others, it's because those stories were delivered by relations who had fled Ireland in the wake of the 1916 Rising or during or after the Civil War. There's a sense of political activism and an innate delineation between 'right' and 'wrong'. For many, that is because of a Catholic faith, a faith that is arguably stronger in the US than it is among their countrypeople in Ireland.

The Irish vote and the Catholic vote can become conflated, but the truth is that neither is a monolithic voting bloc, and neither can be counted on for sure by either party.

Congressman Brendan Boyle of Philadelphia: 'If you take individuals who are conscious of their Irish heritage and very

proud of it, most of those will be Catholic. Some who are very
pro-life vote Republican just because of that reason, and some
who are Irish and/or Catholic and care about social justice issues
and treatment of the poor will vote Democratic because of those
issues. And then there'll be some who are of Irish descent and
identify as Catholic but haven't been inside a church in 30 or 40
years and flip-flop their vote.'

Since the Bush–Gore election in 2000, the quarter or so of
American voters who are Catholic have decided every single
presidential election – whoever won the majority of Catholics
won the White House.

Catholics voted for Barack Obama by a narrow margin twice,
for George W. Bush narrowly twice and for Bill Clinton twice,
and in 2016 Donald Trump won 52 per cent of the Catholic vote.
Indeed, like the Irish American vote, the Catholic vote comprises
diverse political views and a diverse socioeconomic makeup.

An interesting study of how the power of the Irish connection
is viewed by Irish Americans, or just Americans, is to examine the
approach taken to exploiting, or embracing, their Irish heritage
by two of the Democratic candidates hoping to be selected as the
party's nominee to contest the presidential election in 2020. On
one hand is former vice president Joe Biden. He speaks of Ireland
and his ancestors frequently. He hails from Scranton, Pennsylvania,
where a large percentage of the population also shares his Irish
background. He visited his ancestral homeplaces in Ireland while
vice president and is happy to joke about drinking pints of beer
and wearing green ties. He often cites Irish poets in speeches.

On the other hand is Beto O'Rourke, a former congressman
from El Paso, Texas. He is some 30 years Biden's junior, so
perhaps a generational shift is in evidence here. Beto's full name
is Robert O'Rourke, but he uses the common Mexican nickname
of Beto, which is unquestionably less Irish and more Hispanic
– a constituency he was keen to court in his unsuccessful 2018

Senate bid against Ted Cruz. (Cruz has Irish roots on his mother's side and frequently speaks of them.) Although El Paso is right on the Texas–Mexico border, it is actually a city with strong Irish ties, and O'Rourke's own family has a long association with the community. His father, Patrick Francis O'Rourke, a county judge, once ran for Congress as a Republican and was a judge of some note in El Paso.

Joe Biden must consider there to be a national-level value in being associated with Ireland; Beto O'Rourke apparently seems not to consider that to be the case, or at least the Irish factor is secondary for him to the Hispanic one. The Irish American Democrats association has yet to consider who they will back during the Democratic primary campaign this time around. They are going to sit out the fundraising until the field narrows a little.

The Democratic candidate will face the incumbent president, Donald Trump. His team didn't court Irish voters in 2016 – although they received plenty. Nevertheless the President made it his business to stick to a promise he had made to visit Ireland during his first term. He did this in June 2019. Critics argued that the visit was more about promoting images of his golf course and resort in Doonbeg, County Clare, and less about solidifying any affinity with Ireland. However, his campaign team can consider the box ticked regardless.

The president's visit came just several weeks after a congressional delegation led by the Speaker of the US House of Representatives Nancy Pelosi visited Ireland and Northern Ireland, and just before a second congressional delegation came in early July. A visit from Vice President Mike Pence took place in late 2019. Whether any of these high-profile US politicians considered they would score political capital with their electorates by visiting Ireland, they at least considered the relationship valuable enough to make the journey. With so many visits in a short space of time, it is clear the political relationship is alive and well.

Former US ambassador to Ireland Kevin O'Malley is concerned about the future relationship, not merely from the perspective of Irish influence, but also because of the impact that a reduced immigration flow of Irish to the US might have on the United States itself.

'Ireland isn't exporting O'Malleys any more to the US. There isn't much Irish immigration to the US any more. So there won't be people like me growing up. I never thought about Ireland as a foreign country. It was just where everybody came from – that's how I thought about it. So I think it has to be nurtured. I truly do. I don't think it's hard to nurture it. I think the Irish and Americans are leaning towards one another naturally. I don't think it's a big sell that you have to do, but I don't think it's inevitable that this will continue.'

His words are echoed by former congressman Joe Crowley, who lost his seat in November 2018 after 20 years to the rising star Alexandria Ocasio-Cortez, who has no connection to Ireland apart from a shared immigrant background. Crowley's parting words: 'Don't worry. We'll always have your back.'

It also never hurts to have a Plan B.

Irish power and influence in US politics are fading at a domestic political level. Occasionally Irish Americans will rally around a certain candidate, as they did with presidents Kennedy, Reagan and Clinton, but even then, that was for a variety of reasons, not one particular issue, and each time for a different set of reasons. Kennedy was the first Irish Catholic, Reagan promised economic improvements and Clinton promised peace. There is no evidence that a similar performance could be repeated.

Irish political power in the United States is at its most potent when there is a cause for Irish America to rally around. Irish foreign policy regarding the United States relies on an allied relationship: a sense that a friendly yet powerful nation, the might of Irish America, is always waiting in the wings to come to the aid of Ireland if and when it is needed. In recent decades the clearest example of that was the peace process and lead-up to the Good Friday Agreement. Irish American politicians had a cause they could engage with and campaign on. The level of engagement can be seen to fade when there is not a clear reason to 'help' Ireland, and re-ignite when there is, as is the case currently with the uncertainty surrounding Brexit. Who could have predicted that the issue to reinvigorate Irish America would be the decline of Ireland's relationship with Britain? The connections need to be protected during the times when Ireland is self-sufficient and productive and does not need hand-outs, protection or back-up. A relationship that lies quietly yet reliably in the background, unsullied by partisanship, greater than the leaders of the day in both countries.

—

The future of Irish influence in the United States lies in having a multifaceted approach. Hooking people in with music and culture and art, tying them with emotional and cellular-level connections, giving them a tribe to belong to and identify with are still powerful persuaders. But these need to sit alongside new approaches that include economic arguments, two-way investment and education opportunities, both for younger generations of Irish Americans and for those who have no ancestral connections. Any relationship is a living thing that must be nurtured and cultivated. The Irish–American one is no different. It cannot be taken for granted. It is changing, and it will wane and fade if neglected over a long period of time. However, as long as the change is fostered and Ireland plays an active role in shaping it, the future can be red, white and blue, and green – if perhaps a paler shade. Meanwhile, work on developing a Plan B must continue.

IRISH AMERICAN HERITAGE MONTH, 2019

By the President of the United States of America

A Proclamation

During Irish American Heritage Month, we celebrate the indispensable contributions Irish-Americans have made to every chapter of our Nation's history. Generations of Irish immigrants have carried to our shores character, culture, and values that continue to play pivotal roles in the strength and success of America.

Irish-Americans helped define and defend our great Nation in its earliest days. The Continental Congress appointed more than 20 generals of Irish descent to lead the Continental Army through the Revolutionary War. The courage of these Irish generals on the battlefield was as inspiring as it was fierce. The Pennsylvania Line, the backbone of George Washington's Army and one of its largest and hardest-hitting units, consisted of so many soldiers of Irish descent that it was often called the 'Line of Ireland.' After they fought for our Independence, Irish-Americans helped enshrine the visionary principles of self-government outlined in the Declaration of Independence and the Constitution.

Many Irish-Americans immigrated to the United States during the terrible years of Ireland's Great Famine in the middle of the 19th century. Despite facing discrimination and poverty, Irish-Americans persevered thanks to their industry, leadership, and integral involvement in society. In 1868, Irish-American businessman Edmund McIlhenny grew his first commercial crop of peppers in Avery Island, Louisiana, and created 'Tabasco' hot

e. Andrew Mellon, the grandson of Irish immigrants, built nriving business empire before becoming the Secretary of the reasury, during which time he advocated for economic policies that sparked the tremendous prosperity of the 1920s. In 1937, he funded the construction of the National Gallery of Art and donated his extensive art collection to the museum.

Today, more than 31 million Americans look back with pride on their Irish heritage and the legacy of their ancestors. The faith, perseverance, and spirit of Irish-Americans across our country is indelibly woven into the tapestry of the American story. As we spend this month honoring the incredible history of Irish-Americans, especially on St. Patrick's Day, we look forward to a bright future of continued friendship and cooperation between the United States and Ireland.

President Donald J. Trump.

ACKNOWLEDGEMENTS

There may be a hundred thousand welcomes in Ireland, but the same is true in Irish America. Being Irish opens so many doors in the US. To all those who opened their door to me, in the four years I lived in Washington DC and subsequently during the research period for this book, I am truly grateful. There were far more interviewees, far more stories and far more viewpoints than I could fit in this book. To all those who spoke to me, on and off the record, thank you for your time and hospitality, and your candid opinions. Many volumes could be written on the state of play of Irish power and influence in US politics, but I have chosen to keep the focus narrow, to offer a snapshot of where we are now and what the future might bring.

Deirdre Nolan, Sheila Armstrong, Teresa Daly, Avril Cannon, Paul Neilan and all at Gill Books. Marianne Gunn O'Connor, Niamh Tyndall, Noel Kelly, Niamh McCormack and Andy McAnally. Friends and colleagues at RTÉ News, particularly Jon Williams, Hilary McGouran and Keelin Shanley. Friends on both sides of the Atlantic, including those in the Irish American community – a nexus without limitations. My parents, sisters and in-laws and our own expanding family networks. Finally, travelling across the US and writing this book while heavily pregnant would not have been possible without the support of my husband, so my never-ending thanks to him.